AF708952

LONG-TERM AND DYNAMICAL ASPECTS OF INFORMATION SECURITY: EMERGING TRENDS IN INFORMATION AND COMMUNICATION SECURITY

LONG-TERM AND DYNAMICAL ASPECTS OF INFORMATION SECURITY: EMERGING TRENDS IN INFORMATION AND COMMUNICATION SECURITY

ANDREAS U. SCHMIDT
MICHAEL KREUTZER
AND
RAFAEL ACCORSI
EDITORS

Nova Science Publishers, Inc.
New York

Copyright © 2007 by Nova Science Publishers, Inc.

All rights reserved. No part of this book may be reproduced, stored in a retrieval system or transmitted in any form or by any means: electronic, electrostatic, magnetic, tape, mechanical photocopying, recording or otherwise without the written permission of the Publisher.

For permission to use material from this book please contact us:
Telephone 631-231-7269; Fax 631-231-8175
Web Site: http://www.novapublishers.com

NOTICE TO THE READER

The Publisher has taken reasonable care in the preparation of this book, but makes no expressed or implied warranty of any kind and assumes no responsibility for any errors or omissions. No liability is assumed for incidental or consequential damages in connection with or arising out of information contained in this book. The Publisher shall not be liable for any special, consequential, or exemplary damages resulting, in whole or in part, from the readers' use of, or reliance upon, this material. Any parts of this book based on government reports are so indicated and copyright is claimed for those parts to the extent applicable to compilations of such works.

Independent verification should be sought for any data, advice or recommendations contained in this book. In addition, no responsibility is assumed by the publisher for any injury and/or damage to persons or property arising from any methods, products, instructions, ideas or otherwise contained in this publication.

This publication is designed to provide accurate and authoritative information with regard to the subject matter covered herein. It is sold with the clear understanding that the Publisher is not engaged in rendering legal or any other professional services. If legal or any other expert assistance is required, the services of a competent person should be sought. FROM A DECLARATION OF PARTICIPANTS JOINTLY ADOPTED BY A COMMITTEE OF THE AMERICAN BAR ASSOCIATION AND A COMMITTEE OF PUBLISHERS.

LIBRARY OF CONGRESS CATALOGING-IN-PUBLICATION DATA

Schmidt, Andreas U.
Long-term and dynamical aspects of information security : emerging trends in information and communication security / Andreas U. Schmidt, Rafael Accorsi, Michael Kreutzer.
p. cm.
Includes bibliographical references and index.
ISBN-13: 978-1-60021-912-2 (hardcover : alk. paper)
ISBN-10: 1-60021-912-8 (hardcover : alk. paper)
1. Computer security. 2. Internet--Security measures. 3. Telecommunication--Security measures. I. Accorsi, Rafael. II. Kreutzer, Michael. III. Title.
QA76.9.A25S345155 2008
005.8--dc22 2007030778

Published by Nova Science Publishers, Inc. ✢ New York

CONTENTS

PREFACE

Andreas U. Schmidt, Michael Kreutzer and Rafael Accorsi

This book presents selected contributions from the workshops "Security in Autonomous Systems" and "Long-lasting Security" which took place at the conference ETRICS — Emerging Trends in Computer Security, on the 8th and 9th June 2006 at the University of Freiburg im Breisgau, Germany. According to the ambivalence of dynamic and static qualities, the contributions in this volume are separated into two main parts. This major separation is cut along the application dimension of the technology treated, i.e., is the *real-world effect* a *dynamic* or a *static* one? The first six papers deal with technology that operates in rapidly changing environments and has to adapt to their shifting conditions. The technology or application envisaged in the second six operates or assumes a static outcome that can, for instance, be used forensically.

The technological environments of the first, *"dynamic"*, part of contributions are heterogeneous, rapidly changing, where "change" is a consequence of the inherent dynamics caused by:

- *Openness*. Unlike traditional systems, dynamic systems are not restricted to a previously registered set of participants. Participants — and thus the devices they may own or carry along — are able to join and leave the system at their discretion.
- *Adaptation*. As the composition of the system changes, systems adapt themselves autonomously to provide the needed functionality according to the preferences of the new participants and the capabilities of the systems.
- *Emergent behaviour*. As a result of or due to adaptation, new, unforeseen behaviour emerges, e.g., for proactive services based on real-time context information.

Security is challenging due to these attributes. There are various, essentially equivalent, denotations for highly dynamic systems in use today. Authors in this volume use "ubiquitous, pervasive computing" and "ambient intelligence" interchangeably when referring to dynamic systems.

On the other hand, there is a need for *long lasting* security in order to cope with the challenges of real-world effects with *static* outcomes that leave, for instance, a persistent trail for forensic inspection. Long lasting IT-security means that mechanisms guarantee that protection goals will be maintained at least for the maximum lifetime of a human being, i.e., 110 years or even longer. At first sight this condition might sound strange in industrialised societies in which economic processes are constantly accelerating ("ICT enabled real time business") and in which thinking is restricted to time horizons like the present year, or even only the next quarter (share holder value), or at best to the next five years (election periods and research/project timescale of funding bodies and enterprises). However, taking the notion of long lasting security for granted means protecting values at least as long as individuals live. But, can we give at least one example for data that is worth being protected during a lifetime? The answer is yes. Let us take privacy as an example, in particular medical records and here the genome information. This information does not change during lifetime and is potentially of high value. For instance genome information can help prevent complications during a cancer therapy. However, genome information must be handled confidentially. At this very moment, genome information is floating (though not on the level of individuals) through our main open network, the Internet. What happens to people with a "bad genome prophecy" when everyone can decode the IP packets with this payload? Another example for life long confidentiality: The advent of ubiquitous computing with its pervasive tracking technologies challenges the privacy of location information. What will happen if virtually everyone could gain access to location records and can find out where you have been at an arbitrary time in the past?

Not only confidentiality has long term aspects, but also integrity and authenticity. What happens when the integrity and authenticity of data digitally signed by notaries (realty contracts, testimonies, etc.) fails? Can we then determine to whom a house, a piece of real estate or a yacht belongs, without doubt? Here we face an even longer time perspective than the span of lifetime if we take the interests of inheritors into account. Legal regulations in many countries foresee indefinite compulsory periods of record-keeping for such documents.

Concerning the availability of long term mechanisms for these challenges we quote a statement from the Recommendations of the Advisory Board installed as an outcome of the European Research project SecurIST[1]

> Long term security and dependability: Security and dependability issues typically go along with the life cycle of a technology. The trend to first deploy a technology and later fix its problems – typically driven by economic motives – is gradually making way for security by design, resulting in improved security at the beginning of the life cycle. Unfortunately, the security issues of a technology near the end of its lifetime are typically overlooked. The best known example is that of cryptographic keys and algorithms which may need to offer in some cases security for 50 to 100 years. However, this problem also arises in other areas such as electronic documents. Will it be possible to view current documents 50 years from now and, if so, will it be possible to assert their integrity? Many applications stay in use for much longer than anticipated, but during the extended lifetime they will be functioning in an environment for which they

[1] http://www.securitytaskforce.org/dmdocuments/securist_ab_recommendations_issue_v3_0.pdf

> have not been designed, resulting in completely new vulnerabilities and risks. [...] In view of the complexity of the challenging security and dependability problems we are facing today, addressing these issues seems to be far beyond the state of the art.

The articles of the second part of this book try to tackle some of these problems.

A proposal to quantify the time scales on which IT security lives may be a major contribution of this volume. We identify three main axes. First, the *time scale of practical relevance*. It denominates approximately the middle of the time range during which the problem treated by the technology becomes imminent to the IS&T community, through the demands of the market (Yes, the market is the empirical substrate of the applied science called IS&T, and it is the place to validate its theoretical, or lab results). Second, the technological *research and development (R&D) time scale*. It is a forecast of when the technological solution will be "there". This means in effect the first of two milestones in technology evolution: a) when a technology has made the breakthrough to be in principle, i.e., theoretically and economically applicable and viable, and b) when it is matured to reach prevalence in IS&T, i.e., to become textbook standard. The second point in time hinges obviously on external factors not amenable to analysis within the discipline proper. Third, we have the *event horizon* of the technology in question. That deals with the issue of the sustainability of the solution: May it be surpassed by the skills of attackers using advances in other areas? Can it be foreseen that it will be overcome by even better solutions?

Valid as these three categories might be, their strict quantification faces fundamental problems, or rather, is tantamount to clairvoyance. The problems are similar to those faced by the authors of the numerous "technology road maps", which are so desired as a guidance tool by policymakers and funding bodies. Let us be pragmatic and take an "everybody does it and so do we" approach and see, what comes out when we try to order the twelve articles at hand. The outcome is contained in the table below. The reader shall be our judge.

Let us now come to the individual contributions in this volume. Usability is – at the first sight at least – not related to security and privacy. However, this is deceiving. Up to 75% of the total amount of user errors are security relevant. In dynamic environments this is aggravated, as most interactions are implicit in the sense that individuals are unaware of them. Individuals eventually end up losing the control over the disclosure and usage of their personal attributes. Empirical investigations show that the potential of implicit interactions may also have a negative impact on the acceptance of dynamic IT systems. Recognising this, Höhn et al. propose in the first article an innovative approach to visualise implicit interactions in them. Unlike other proposals on this topic, the visualisation does not happen during the interaction but instead afterwards, using views on the collected data.

Visualisation techniques display a significant improvement regarding privacy, still they keep individuals out of the loop, as they cannot take appropriate actions to tell the system how to handle their interactions. To this end, security and privacy policies are of utmost relevance, for they provide users with a language to express their preferences; monitors based on different trust models enforce these policies. Contribution no. 2 of Kähmer identifies missing elements of the policy language EPAL used by IBM and extends it by operators to simplify the negotiation of policies between individuals and the system. In the same spirit, Hutter focuses in paper no. 3 on a particular kind of policy, namely information flow policies, for the privacy-aware composition of web services. He defines a formal framework

Table 1. The twelve articles in this volume ordered along the three described axes and the main category. [1]Means that the research is of fundamental character viz. the technology basic, and there is no foreseeable solution. [2]The audit trail produced by re-signing arguably loses probative force on historic time scales. [3]Until a stable jurisdiction has formed.

No.	Short title	dynamic vs. static	practical relevance time	R&D time	event horizon
1	Secure Usability for Ambient Intelligence	dynamic	now	now	fundamental n/a[1]
2	Privacy Policies for Dynamic Systems	dynamic	now $\sim\infty$	now	fundamental n/a[1]
3	Web Privacy by Information Flow Control	dynamic	now $\sim$5 yrs	now	fundamental n/a[1]
4	Secure Routing for Mobile Ad Hoc Networks	dynamic	now $\sim$5 yrs	now	fundamental n/a[1]
5	Pervasive Access Control Architecture	dynamic	now $\sim$5–8 yrs	now	fundamental n/a[1]
6	Real-time Watermarking for Long-term video preservation	dynamic	now $\sim$10 yrs	now	watermarks broken?
7	Long-term Secure Key Establishment	static	$\geq$15 yrs	now $\sim$3–5 yrs	potentially ∞
8	Quantum Cryptography for Mobile Networks	static	3–5 yrs	now $\sim$20 yrs	∞ by physics
9	Legal Concepts for Long Term Validity of Digital Signatures	static	now	now	depends on legal framework
10	Long-term security for signed documents	static	5–8 yrs	now	$\sim$ 100 yrs[2]
11	Providing Evidence with Transformed Signed Documents	static	5–8 yrs	now $\sim$5–8 yrs	$\geq$ 100 yrs[3]
12	Design Strategies for Secure Embedded Networking	static	now $\sim\geq$30 yrs	now	preliminary approach

for the definition of web services, their canonical composition into web spaces and proves that the corresponding composition is secure with respect to information flow policies.

Web services live on a high abstraction layer and attacks on lower layers can still compromise the system as a whole. For example, routing attacks in which the data flow between two recipients is re-routed by an attacker via a station under his/her control. Ad hoc wireless networks are a particularly convenient environment for such attacks. To tackle this problem, König et al. introduce a novel paradigm for geographically secure routing that constrains the flow of information to the trustworthy nodes in the fourth paper. To this end, the authors pursue a cross-layer approach to couple the application layer – higher level security requirements – with the network layer, i.e., usual routing protocols.

Although a plenty of techniques and security mechanisms for dynamic systems can be found in the literature, there is a noticeable lack of attempts to deploy them in an integrated fashion. The contribution no. 5 of Anisetti et al. focuses on the development of OpenAmbient, an architecture based on web services that incorporates an access control language allowing the expression of ambient predicates, i.e., context information.

While the other contributions in this part focused on fundamental research needed to deploy dynamic systems, contribution no. 6 of Yamada et al. closes the first part of this volume by providing and evaluating a architecture for real-time watermarking of dynamic content, e.g., movies or other broadcast media. Applications of this kind are of gaining visible relevance as a way to protect proprietary content, as watermarking presents a more flexible approach as, e.g., digital rights management.

The first article of the second part by Bohli et al. describes a main threat to classical crypto mechanisms and in consequence to long term security, namely the advent of the quantum computer. Based on the different approaches to build such a new type of computer and their success stories the threat to the secure establishment of cryptographic keys they pose might become imminent within the next 15 years. In view of this threat the authors present a solution which in principle could be implemented today (or in not more than 3–5 years) based on light assumptions (the quantum computers cannot break symmetric encryption and the Diffie-Hellmann assumption holds).

Marhöfer et al. give in paper no. 8 an overview on how quantum cryptography can be used as an opportunity for long term security. The first quantum crypto companies exist and we think that they might make considerable money in 3 to 5 years from now. The fascinating charm of this approach is that security based on natural laws could last as long as these laws are valid, i.e., in principle forever!

The 9th paper by Adamski et al. is a sharp analysis of the problems of wide spread legal concepts for long term validity of digital signatures. The discrepancy between technical reality and the model adopted by legal systems is a current problem which should and could be solved today. Depending on the governing law this problem will show up very soon in different nations.

Thomas Kunz et al. present a technical solution for how long-term security for signed documents can be achieved. The key length or even algorithms of signing lead to insufficient security of digital signatures over the course of time. This problem that will be here in not more than eight years. It must be taken into consideration that even with the elaborated solution presented in the paper (which may be used for 100 years) the audit trail produced by re-signing may lose probative force on historic time scales. Wilke looks at the same problem from a legal perspective in paper no. 11, and points to shortcoming of legal regulations in Germany. As signing practices of the past will be insecure in at most eight years we need such a legal framework until then. This framework might be changed and adapted to the advent in technology but the kernel of its idea might last as long as we live.

The last article of this book by Sikora and Braun emphasises the fact that more and more embedded systems are and will be deployed. Many of them run over decades but cannot be changed. The presented solution to this long-term change management security problem is an interesting preliminary approach and its further development might have a high potential in the future.

The treatment of IT systems security on time scales is at its beginning. This book aims at giving an impetus to the formation of the emerging field. We hope the reader will share our enthusiasm for the new paradigm, and maybe even re-think it critically.

Frankfurt am Main, Darmstadt, and Freiburg, April 2007
Andreas U. Schmidt, Michael Kreutzer, and Rafael Accorsi

After studying mathematics and physics and graduating in mathematics in 1994 at the Johann Wolfgang Goethe-University Frankfurt, Germany, Andreas U. Schmidt received his doctorate in mathematics with a dissertation on mathematical physics in 1999. After research stays in Durban, South Africa in 2000/01, and in Pisa, Italy, in 2001/02, he returned to the Fraunhofer Institute for Secure Information Technology SIT in Darmstadt in 2002 where he acts as a senior researcher and project manager. His areas of active interest are diverse and include various subjects within applied information security. Andreas is currently the co-ordinator of the project TransiDoc, funded by the German ministry of Labour and Economy, on legally secure transformations of digitally signed documents. He applied TransiDoc concepts to electronic notarisations and authorised translations. His current research foci are: i) theoretical questions in information economy with practical implications for the marketing of virtual goods, ii) applications of Trusted Computing technology with focus on the mobile domain, iii) security and electronic signatures for multi-media communication, in particular Voice-over-IP. iv) long-term security of and secure transactions with electronically signed documents. Andreas is co-founder and director of the board of the novalyst IT AG.

Dr. Michael Kreutzer is the Managing Director of the Darmstadt Centre for IT Security (DZI) at the Technische Universität Darmstadt. After completing his degree in Computer Science at Saarland University, he began working at a Luxembourgian company on a project dealing with multilingual search engines for the EU legislature. Michael then moved to a renowned consulting firm for distributed object computing. In 1999, he won first prize at the OMG Awards in Berlin for an innovative, object-oriented banking solution. During his time at the University of Freiburg, he published solutions for the security and privacy problems that accompany the use of mobile computers and contributed to visionary projects such as "Living in a smart environment", with the Gottlieb-Daimler- and Karl Benz-Foundation. Michael is co-author of the textbook "Telematik- und Kommunikationssysteme in der vernetzten Wirtschaft". In July 2006, he was awarded his Ph.D for his dissertation on scalable and robust service discovery in mobile ad-hoc networks ("Skalierbare und robuste Dienstfindung in mobilen Ad-hoc-Netzen"). Since 2005, he has worked in Darmstadt, the research capital for IT Security in Germany. At the Darmstadt Centre for IT Security, Dr. Kreutzer is pursuing the goal of bundling the myriad of research activities in the field of IT Security in Darmstadt and to make Darmstadt's IT Security expertise visible, both nationally and internationally. One of the DZI's core competencies is long-term security.

Rafael Accorsi is a research assistant concluding his Ph.D at the University of Freiburg, Germany. He obtained his BsC. in Computer Science at the University of Pelotas, Brazil, in 1998 and completed his MsC in 2000 in Logic and Computer Science at the University of Amsterdam. Rafael moved to Freiburg to work with David Basin in the formal verification of security protocols. At concluding the project, Rafael started in 2004 at the Department of Telematics of the University of Freiburg, where he carries out research into privacy and security in highly dynamic systems. His upcoming dissertation focusses on the automated audits to investigate violations of security in dynamic systems. His current research interests include the development of policy languages, formal expression of security properties, secure logging services and auditability of dynamic systems.

In: Long-Term and Dynamical Aspects of Information... ISBN 1-60021-912-8
Editors: Andreas U. Schmidt et al, pp. 1-17 © 2007 Nova Science Publishers, Inc.

Chapter 1

AN APPROACH FOR SECURE USABILITY FOR AMBIENT INTELLIGENCE ENVIRONMENTS

***Sebastian Höhn*[*]*, Rafael Accorsi*[†] *and Mathias Maier*[‡]**
University of Freiburg, Dept. of Telematics
Friedrichstr. 50, 79098 Freiburg, Germany

Abstract

Many transactions of our daily lives depend on computer systems. This means that newly deployed systems must be carefully designed in order to be correctly used by laypersons. The scenario we present encompasses a large number of sensors in a shopping environment. Since these environments are publicly accessible and the sensors are able to collect data without the users' notice and hence without their explicit consent, privacy issues play an important role. Because it is infeasible to prevent these highly dynamic systems from collecting data, we propose transparency for data collection, storage and processing as a possible solution.

Reliable data of the systems' behavior are an essential part of an architecture for transparency. To this end, secure logging mechanisms are of foremost importance. Furthermore, it is still an open question how this transparency can be presented to the users in a simple and intuitive fashion.

1. Introduction

Since computer systems became networked, dynamic and embedded in everyday objects, the complexity of administration and usage has exponentially increased. Together with this evolution, we are losing the ability to protect our privacy while interacting (perhaps even implicitly) with modern ubiquitous computing systems. We are obliged to deprive ourselves of the possibility to choose whether or not to disclose certain information (e.g. information about our whereabouts) as computer networks collectively and invisibly collect data with their various sensors.

---

*E-mail address: hoehn@iig.uni-freiburg.de
†E-mail address: accorsi@iig.uni-freiburg.de
‡E-mail address: mathias.maier@gmx.net

At present the number of environments where ambient intelligence environments track humans and dynamically interact with their users is still small. Nevertheless these systems are entering our lives in many different areas. Several application scenarios show that all the technologies are at our hands and that they can readily be combined to provide new services and additional assistance. Among the frequently cited scenarios are personal shopping assistants, home automation scenarios and assisted living for elderly or impaired people [12, 5, 17].

Providers of personal shopping assistants claim that customers' acceptance of the involved technology is good [25], although they had to face calls for boycott from privacy activists and organizations when a hidden RFID-chip was discovered in the customer card [16]. This shows that systems of this new type have to be designed with special care for security and privacy in mind in order to be useful and usable to users without background knowledge and without being enigmatic and therefore frustrating.

In this article we first present a scenario involving a personal shopping assistant based on ambient intelligence technologies. We point out privacy issues which arise from collecting data invisibly and without the customers' consent and which counteract the otherwise great opportunities for the retailers' customer relationship management. We then propose the usage of *log views* to create more transparency as a means to lessen the customers' frustration by the ambient systems when they act in an unexpected way or when they violate the customers' privacy policies. Log views provide the customers with evidence which has by definition to be unaltered. To this end, we deliver the necessary technological tool of *secure logging*, which we describe in detail in chapter 4. We conclude with reasoning why transparency will lead to a win-win situation in the privacy domain for retailers and customers.

2. Scenario

To visualize the technologies and processes involved in our scenario we present a short introduction to the supermarket of the future. Some other scenarios are presented in [3].

Because Alice is frequently visiting this supermarket, she owns a combined customer and rebate card. During her visits to the store she is interested in the convenience provided by the innovative technological concepts. She is not really interested in the security and privacy enhancing technologies deployed in the super market. This behavior is common to 85% of the people using the new technologies [15].

Bob, on the other hand, cares a lot for his privacy preferences and thus, refuses to get the customer card. He is not interested in sharing his personal information for a small rebate.

When entering the shop both, Alice and Bob, get themselves a shopping cart with a personal shopping assistant attached to it. The shopping assistant is a computer equipped with a touchscreen, RFID and barcode scanners. It is linked to the backstore network via an integrated wireless network module, and the location of the cart can be tracked at any time to provide the customers navigation information within the store. As neither Alice nor Bob carry their own devices with them, they do not connect to the assistants open interface via near field communication. This interface enables the transmission of shopping lists or special offers directly to the shopping assistant. Near field communication interfaces are also used to replace customer cards with the users' devices and for a new mobile payment system [22].

Alice presents her customer card to the shopping assistant and is consequently provided with the shopping list and some personalized special offers. She has uploaded the shopping list to her personal page on the store's Internet portal.

Bob enters his shopping list directly into the assistant's interface. He will do this, because he is not familiar with the store and wants the shopping-cart to guide him to the products on the list as quickly as possible.

While the electronic shopping list is a nice feature for regular customers, and a help for one-time customers in finding products faster, it is a very powerful instrument for the retailer. The electronic shopping list enables the analysis of the products that each of the customers intended to buy when entering the store, and which products they bought in addition. Together with the possibility of being able to locate the customer in the store via the shopping cart, and the resulting possibility of personalized marketing with info-displays on the shelves, it is possible for the CRM-system to measure the effectiveness of different kinds of marketing techniques on any particular customer. Furthermore, it is possible to dynamically modify marketing strategies in the customers' direct surroundings in order to positively influence their inclination to buy.

Alice is standing in front of the dairy products now, and since she regularly buys the same brand of yogurt, the Electronic Shelf Labeling which is connected to the store's central pricing system updates the label and displays a discounted price for her. She happily takes a few cups of yogurt and scans them.

Bob observed Alice buying the seemingly cheap yogurt and decides to buy some cups of it, too. When he scans the yogurts with his shopping assistant however, that shows the regular (and therefore higher) price, as does the label on the shelf now.

Alice knows that she has earned the right to buy yogurt for a discounted price, and enjoys the effect of the labels "magically" presenting the price of the current special offer. Bob does not know this and is annoyed, because he has to pay more for the same product in the presumably same situation. Things happening without apparent reason are very welcome, if beneficial. However, it is also very frustrating if they show a discriminating nature [23]. It is strictly necessary to help customers understand why someone gets discounted prices on certain goods while others do not.

Alice moves on to the frozen meat. The info terminals keep flashing advertisements for vegetarian products at her. They do that ever since she has experimented with a vegetarian diet. It annoys her to be reminded of this experience now that she enjoys eating meat again.

Generating profiles of customers and classifying them into groups for marketing is sensible from the retailer's point of view. Yet it is a tricky business since it invades the customers' privacy and over-fitting is prohibitive. In this case, Alice was classified as a vegetarian on mistake.

Bob has finished his shopping. However, he finds that he does not want to buy the yogurt any more now that he knows he has to pay more than he thought at first. Being annoyed he doesn't bother to return to the shelf with the dairy products and puts his yogurt on the cheese shelf.

The shelves constantly scan the RFID-chips of the products to report shortages and misplaced products. The computer system in the shop can relate a misplaced product to the customer standing in front of the shelf with his cart or if Bob has already scanned the product for the automated checkout procedure directly with his shopping cart. This could be abused by the retailer (for example to punish customers for undesirable actions). Although Bob thinks that he is anonymous, the system could have analyzed his movements in the shop and the items he bought over time. Based on these statistics the system could single out his individual patterns as matching to some customer who regularly misplaces products. The system cannot make a correct decision whether it actually was Bob, or whether he was unlucky to coincidentally match the movements of a customer notoriously misplacing products.

Bob proceeds to the checkout. He hands over the shopping assistant to the cashier. He is lucky that he does not have to place the contents of his shopping cart on the conveyor belt. There are random tests once in a while to ensure that all the contents was scanned by the customer and therefore registered with the shopping assistant. Before leaving the store, he passes a gate which deactivates anti-theft protection.

Alice checks out, too. She pays with her cash card and takes the detour to the RFID deactivator. In order to erase the data from the RFID chips she has to place each individual tagged product on the deactivator and press a button.

It is not possible for the customers to directly perceive if an RFID chip is readable or which information it contains. Neither is this possible for anti-theft protection. While for the latter non-observability is essential, it is a problem in the former: Alice does not want to carry tagged products out of the shop. She has to trust the retailer that the deactivator works correctly and that it is not manipulated. She would lose her trust in it immediately, if she heard of any malfunction of the device. If she was able to perceive the state of the RFID, she would not need to trust anything or anyone.

3. Observability as a Basis for Usability

The scenario presented in the previous chapter illustrates the characteristics of such highly dynamic systems in an informal way. In the following chapter we will take a closer look at the characteristics of the ambient intelligence environments:

1. These systems are *open*, i.e. participants are able to join and leave the systems at their will.

2. They are *context sensitive*, i.e. they continually adapt to the special requirements of the participants and the technical necessities.

3. They have *autonomous properties*, i.e. there is a vast amount of automated tasks in order to keep these systems in a sane state

4. They are *invisible*, i.e. the users are not directly involved in observing and controlling the systems.

If we summarize these properties and consider the challenges for privacy and usability, we realize that all of these contribute to a *loss of control* realized by the users.

Users cannot observe who is participating in the different stages of communication and to whom information is transmitted. Data are sent over different networks and services of a plethora of providers are combined in order to establish a highly dynamic system. In the supermarket scenario above not all the services will be provided by the owner of the shop. The owners certainly will add new services from specialized service providers, for example, to offer deferred payments. This shows that there exists no well defined area of trust within these systems.

The system's reaction to context makes the interactions between users and systems implicit. These implicit interaction models act like a two-edged sword: on the one hand, they facilitate the users the performance of their tasks and the achievement of their goals. On the other hand, they are not very well suited to support direct control by the users. It is a proactive interaction model that must correctly anticipate the users' intentions. But how can the users' desires correctly be anticipated? And how can the users *observe* what has been anticipated and what decisions have been made on their behalf? If this anticipation fails and the users can realize this failure they become annoyed and directly feel the loss of control.

The same is true for autonomous properties. They are necessary in order to enable such a complex technical infrastructure, but failure of these mechanisms will not be accepted by the users as well as the administrators, in this case. Perhaps the most important aspect for end users is the invisibility of many of the devices. RFID tags are integrated into the packing of the products, but they cannot easily be recognized. Even if they are supposed to be marked, one must trust the producers that they really marked their tags correctly.

Taking stock, it is essential to provide proper mechanisms in order to bring the user back into control. How can we achieve this? If we take into account the seven stages of action presented by Norman [18], we realize that it does not suffice to provide means of control, such as a kill function on the RFID tag. It is furthermore necessary to provide adequate observation mechanisms, too. People who do not trust the provider and hence wish to disable their tags will not trust a fully automated and hence invisible and unobservable deactivation process. They must be able to observe the outcome in a reliable and trustworthy manner.

Unfortunately it is not that simple. Beside the system and its single devices the users have context, too. Depending on their current situation they cannot operate special devices or receive special messages. This is most obvious in systems that are deployed in environments that require the users' full attention, for example in cars or in the working environment. But it is of course also of great importance in the scenario presented above. If the users are carrying some goods they cannot operate their devices, if they are at the check out they will not be interested in goods they might receive at the bargain-basement. This shows that it is important to correctly adapt the user interfaces to the current states of the users and their current preferences, where these preferences cannot be seen as unalterable over time.

Future paradigms for human computer interaction hence must take into account several types of context. Classify this context to tell apart the different states of the devices on the one and the states of the users on the other hand. Furthermore, they must proactively adapt the user interfaces to these states in order to present the information required in a very convenient and unobtrusive way. This means that user interfaces for ambient intelligence

Figure 1. How can the security properties be evaluated?

systems will be decoupled from the actual implementation. Services that make up a system must provide enough information to implement algorithms that can automatically propose the best way to present information and receive input from the users.

4. Unobtrusive Observability

In his seminal book "The psychology of everyday objects" Don Norman presented the seven stages of action [18]. These stages form the process of mostly any action people perform. An elemental part of this theory of action is a feedback loop. That means, people can not plan the accomplishment of very complex tasks in one single step. They, in fact, start performing one subtask they consider correct, observe the outcome of that subtask and then decide on the further steps. To this end, it is necessary that they are able to observe the system's behavior at any time, because the system's state might be an important parameter when planning the next actions.

When designing complex and highly dynamic systems it is very difficult to provide adequate levels of observability. The system's state is most often described in technical terms, such as the type of encryption algorithm used or the length of an encryption key in bits. This seems adequate for the designers and programmers of an application, but it is impossible for laypersons (and often even experts) to decide whether, for example, the given algorithm with the given key length is an adequate choice in the current situation (cf. the interface depicted in Fig. 1).

This becomes even worse in the scenarios of Ambient Intelligence where we have *Mixed-Mode Systems*. This are systems that are build of heterogeneous components that vary in processing power, storage and energy supplies. This worsens the situation for implementing observable security, as additional parameters must be taken into account. Consider, for example, devices with slow processors or a small energy supply that cannot perform the strongest encryption at hand. What kind of encryption is the adequate one in this special case? What happens if we reduce the strength of the encryption in favor of speed? It is beyond the scope of this paper to discuss the issues related with the area of conflict between security and resource poor devices in general, but it is important to realize that for unskilled users this is a problem way beyond their technical understanding.

From the usability point of view there is only one viable solution: system states must be abstracted from their concrete implementation and interpreted in a way the user can comprehend. For the encryption example above the actual estimate of security must be presented not the algorithms. This must be an adequate measure, for example, an estimate of the time an attacker needs to break the encryption. If this is set into relation with the maximum level of security the device can provide and the battery life time, users are able to adjust the device to their needs.

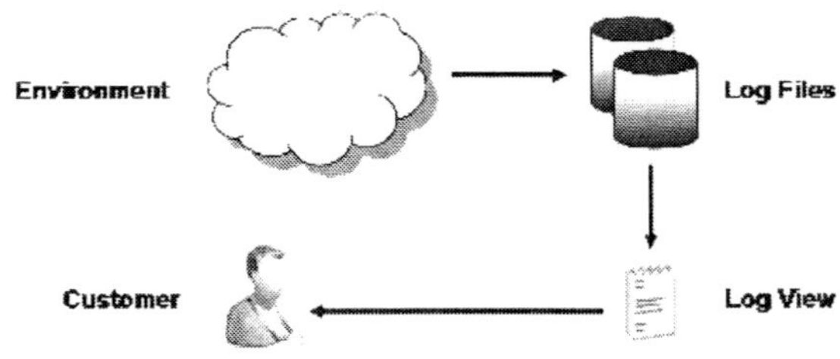

Figure 2. Schematic representation.

As a metaphor one can think about a speedometer of different system properties. For mobile devices battery lifetime under certain assumptions could be one of these properties, but also the aforementioned time needed to break an algorithm with the given key length or the time required to guess a password entered.

It is important to realize that all these quantifications that make up the "speedometer" are just estimates. Currently these estimates are calculated by experts when they plan and implement an application. For future systems that are by definition open and an ad hoc combination of services, this becomes infeasible. As it is impossible to predict the exact composition of a system at a given point in time, it is impossible to specify the system's states in advance. It is hence impossible to provide the correct algorithms and security mechanisms in advance. It is necessary to provide a collection of different mechanisms and algorithms for the user to chose from together with the estimates of their performance (regarding security as well as consumption of resources). If this information is accurately presented, the users are in a position where they can successfully accomplish their tasks.

This does not necessarily mean that the user must decide about every mechanism and algorithm by itself. It is possible that the applications do proactively chose the adequate ones according to the users' policies. These policies must contain the objectives of the users regarding the different measures. Only if it is impossible to find an adequate configuration that satisfies the different objective, it is necessary to bother the user with the different settings of the system.

In the following chapter we will thus provide an approach for log views and secure logging. The secure recording of events within the system is the most fundamental building block for all approaches that require observability. Any visualization of data heavily depends on the quality of the data it needs to visualize. This is overly true in the context of the dynamic and flexible scenarios presented above.

5. Secure Logging and Log Views

Our goal is not to avoid loss of control from the outset, as this would require effective mechanisms to prevent personal data from being collected and regulate the usage of such data, thereby anticipating that customers have complete knowledge about the systems or that their behavior is predictable. This is in conflict with the properties presented in Chapter 3..

Instead, our goal is to provide the customers with evidence regarding how their data has been collected and processed within the system, as schematically depicted in Figure 2. How that information is presented to the users is discussed in Chapter 6.2.. Here, ordinary events happening in the system – e.g. sensor measurements and interactions with RFID-tags – are recorded in (possibly distributed) log files. These log files exist for several reasons. On the one hand, they are indispensable when it comes to managing the systems and pinpointing potential conflicts and failures. On the other hand, they are containers of context data upon which personalization strategies such as individualized services and recommendations could be based upon. Finally, *log views* are the construct we employ to make the behavior of the system evident to customers. Similar to database views, log views encompass all the entries in the log files related to a particular customer. Log views are generated in the system and sent to the customer.

To realize the approach proposed in Figure 2, we must first ensure that collected log data shall not be maliciously modified, i.e. log entries remain authentic throughout the necessary time span. Second, it must be ensured that log views are designed in a way that customers are able to comprehend the information they contain. In this section, we present a secure logging mechanism, as it is a central building block for the generation of log views. For details, see [1, 2].

5.1. An Approach to Secure Remote Logging

Log data can only provide a sound basis for further services when it is authentic. We define authenticity as the simultaneous fulfillment of data *integrity* and *uniqueness*, as illustrated in Fig. 3. *Confidentiality* of log entries is necessary for privacy and is considered as an extra protection goal. A log service is labeled *secure* when integrity, uniqueness and confidentiality properties are fulfilled.

- Integrity states that log data faithfully reflects the state of the devices, i.e., the log data is accurate (entries have not been modified), complete (entries have not been deleted), and compact (entries have not been illegally added to the log file). Thus, log data is not modified, deleted, or appended during the transmission to, and storage at, the collector.

- Uniqueness states that log data shall not allow for parallel realities. Concretely, it is impossible to intercept log data sent from d_1 to c_1 and to resend it (possibly in modified form and claiming a different device identity) to c_2. Log data must be uniquely tagged.

- Confidentiality states that log entries cannot be read by unauthorised individuals, for this would harm inner privacy. Note that confidentiality is also related to uniqueness, for log data transmitted in clear-text can be easily duplicated.

These properties are implemented with cryptographic techniques, which need to ensure *tamper evidence*, i.e., attempts to illicitly manipulate log data must be detectable to a verifier [14], and *forward integrity*, i.e., log data contains sufficient information to confirm or rebuke allegations of log data modification before the moment of the compromise [4].

The goal of the attacker is to gain access to private log data and, thus, to violate its integrity, uniqueness, and confidentiality. The threats posed by an attacker are described using

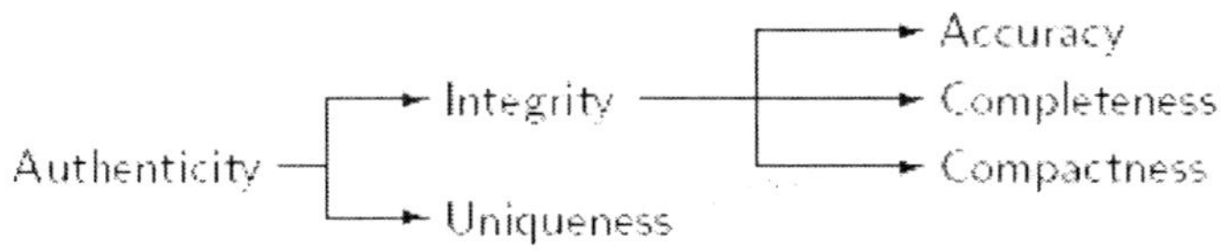

Figure 3. Authenticity property for secure logging.

an attacker model. While we are aware of recent ongoing research on formally characterizing attacker models for dynamic environments [7], we refrain from sticking to a particular model.

5.2. Overview of our Approach

Our approach to secure remote logging services in dynamic systems is based on and extends the techniques proposed in [20]. The idea is to devise a protocol to securely store log data. The protocol starts at the device: it is in charge of applying cryptographic techniques to ensure tamper evidence and forward integrity. When the device runs out of storage, it contacts the collector to request remote storage. The protocol ends with an irrefutable proof of possession of the collector. In detail, our protocol consists of the following main steps:

1. *initialization and construction of the log file*: the device is in charge of applying cryptographic techniques to safeguarding the integrity and uniqueness of its log file. To this end, it computes a secret random value pv_0 and, for each entry, a proof value Z associated to that entry based on pv_0. The Z-values build a chain, so that the integrity of the whole chain can be checked by analysing its last link.

2. *mutual authentication of services*: apart from authentication, the device and the collector also agree on a secret value pv_0 that will be used to ascertain authenticity of log messages.

3. *acknowledgement of receipt from collector*: by receiving the chunk of log data, the collector computes the Z-value associated to each entry and sends the last Z-value signed together with a timestamp and protocol sequence number back to the device. The device then stores this unambiguous piece of information, as it demonstrates that the collector received the chunk of log data correctly and can be thus held accountable for attacks upon this data.

We describe the initialization, appending, and acknowledgement processes below; the authentication phase and the secure exchange of pv_0 are taken for granted.

5.2.1. Initializing the Log File

Assuming that the device d successfully generates a secret value pv_0, d creates the log file by inserting the first entry into it. We illustrate the initialization entry L_0 in Fig. 4, where the fields stand for the following information (we remark that all entries have the same format):

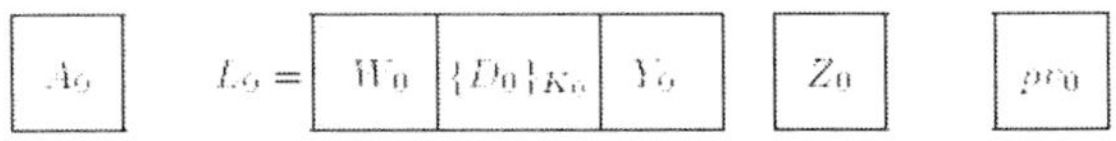

Figure 4. Initial log entry L_0 and values A_0, Z_0, and pv_0.

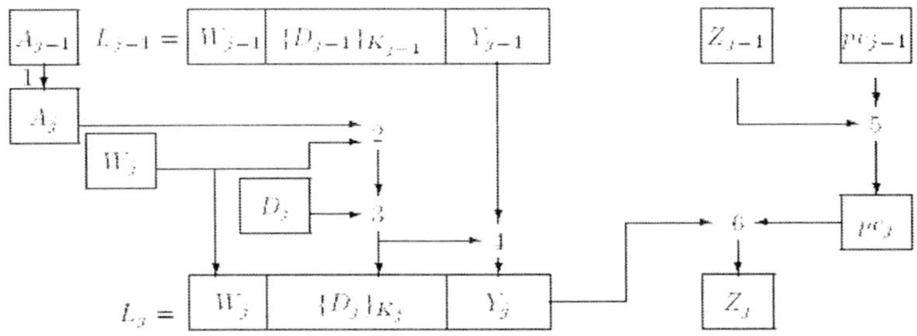

Figure 5. Adding an entry to the log file.

- W_0 is a permission mask to regulate the access to the log entry L_0. According to [20], at the initialisation phase this entry type may be set to **LogfileInitializationType**.

- $\{D_0\}_{K_0}$ is the symmetrically encrypted log data for the entry L_0 and K_0 is a random session key. To provide the necessary security guarantees, D contains not only the event to be logged, but also a timestamp and a protocol identifier. (The former states the actuality of the entry, the latter avoids harmful protocol interactions between different protocol steps.)

- Y_0 is the first link of a hash-chain.[1]

- A_0 is the authentication key for the first entry. As we discuss in [1], this is the secret upon which the security of the whole protocol rests. In practice, since we do not assume that the device is tamper resistant, we suggest that this secret should be kept off-line.

- $L_0 = W_0, \{D_0\}_{K_0}, Y_0$ stands for the initial log entry.

- Z_0 is the message authentication code of L_0 defined as $MAC_{pv_0}(L_0)$. This piece of information is used to compute the proof value associated to the whole chunk of log entries and, thence, can be used as a challenge against the collector. For this, it will *not* be send along with L_0 to the collector.

5.2.2. Appending Log Entries

After creating the log file, the device starts adding entries to the log file. This is schematically illustrated in Fig. 5, where the numbers correspond to the items below.

[1] In the simplest form, a hash-chain Y is inductively defined as $Y_1 = Hash(Y_0)$ and $Y_n = Hash(Y_{n-1})$.

1. $A_j = Hash(A_{j-1})$ denotes the authentication key of the jth log entry. The confidentiality of this information is for accomplishing the aforementioned security properties, as it is used to encrypt log entries. Thus, we assume that the computation of the new value irretrievably overwrites the previous value.

2. $K_j = Hash(W_j, A_j)$ is the cryptographic key with which the jth log entry is encrypted. This key is based on the permission mask W_j, so that only authorized services gain access to the entry.

3. $\{D_j\}_{K_j}$ is the encrypted log entry.

4. $Y_j = Hash(Y_{j-1}, \{D_j\}_{K_j}, W_j)$ is the jth value of the hash-chain. Each link of the hash-chain is based on the corresponding encrypted value of the log data. This ensures that the chain can be verified without the knowledge of the actual log entry.

5. $pv_j = Hash(Z_{j-1}, pv_{j-1})$ is the proof value associated with the jth entry.

6. $Z_j = MAC_{pv_j}(Hash(L_j))$ is the authenticator of the jth log entry. Note that we compute the message authentication code for the whole entry instead of a field of it (in the case of [20], the hash-chain value Y).

The log entry L_j generated by the device consists of the permission mask W_j, the encrypted log entry $\{D_j\}_{K_j}$, and the hash-chain value Y_j; it is denoted by $L_j = W_j, \{D_j\}_{K_j}, Y_j$.

5.2.3. Acknowledgement Phase

Assuming that the authentication between the device and the collector succeeds and they agree on the proof value pv_0, the last phase of the protocol aims to provide irrefutable evidence regarding collector's possession of the chunk of log data sent by the device, as well as the chunk's integrity. To this end, the following steps are carried out:

- by receiving the chunk starting at L_j and ending at L_k (with $j < k$), the collector computes for each entry L_i the corresponding pv_i and, thus, Z_i values.

- after $k - j$ iterations, the collector obtains the authenticator $Z_k = MAC_{pv_k}(Hash(L_k))$.

- the signed proof value is sent to the device $Sign(Z_k)_{K_c^{-1}}$. This message includes a timestamp and protocol step identifier.

- the device then checks whether the authenticator matches with the authenticator computed during the second phase. If it does, the device frees the storage by deleting the chunk.

This completes the presentation of the secure logging mechanism upon which the generation of log vies is based. We refer to [1] for a detailed analysis of the secure properties of this mechanism, in particular the fulfillment of the authenticity properties presented in Chapter 5.1..

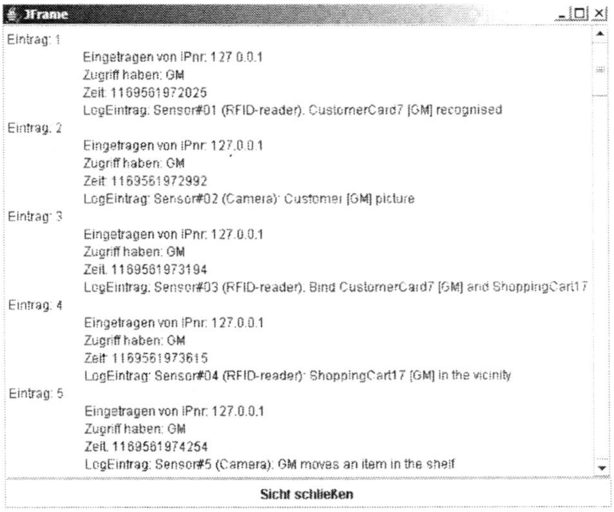

Figure 6. Example of a log view.

5.3. Generating Log Views

Log views are generated by querying possibly distributed log files for a particular individual. Suppose individual *GM* wants to be aware of all the events he or she has been engaged to in the system. To this end, *GM* requests the corresponding log view, i.e. a logical table consisting of the entries of the log file.

To generate the log this end, we employ the access control field W of each entry as an index. The individual – *GM* according to our running example – authenticates to the system and queries the log files for entries whose index, i.e. $W = GM$, matches to the authenticated individual.

The generation of view proceeds without further ado in case the events triggered can be related to an individual. Figure 6 depicts a view for individual *GM*.

At the moment, we consider only the case where individuals are identified at the moment of the event collection. However, should this relationship not be known in advance, techniques such as data mining or sensor fusion should be applied to make the relationship evident, at least up to some statistical probability. Our current research focuses on this issue.

6. Usable Security and Privacy for the Future Super Market

How can we integrate these mechanisms to provide usable security and privacy for end users? Based on the classification of services according to the amount of private data they use we propose the integration of this information in contextualized prototypes for the end user.

6.1. Individualized and Personalized Services

In [19] we presented a classification of services according to the need of input data necessary to provide the service. This classification is well suited for the economical perspective the provider takes. But it is very misleading if we take the users' perspective. It is important to realize that this static view we presented does not hold for the users moving around in the area controlled by the provider. The longer they interact with the system, the more information aggregates with them. This information is not vanishing over time, that means that the users in this categorization will inevitably drift from universal ones to personalized ones. If they interact with the system long enough, it will be possible for the provider to personally identify each of them.

For the users shopping in this super market it is important to know at which level they actually interact with the system. Before the first interaction they are certainly unidentified. But when will they be uniquely identified? We propose the implementation of a "anonymeter" or "anony-lights" in order to present the user this information. As a measurement we evaluate the securely logged data about the interaction with the system and try to find the user's profile. As long as there is no interaction with the system, we can not aggregate more information than all the user profiles have. Thus the probability to get the user profile is as high as randomly picking one of the profiles:

$$P_{ident} = \frac{1}{\text{number of profiles}}$$

While the users interact with the system we get some of their attributes and can sort out some of the users that do not match with these properties. As a consequence the probability of uniquely identifying the user increases (i.e. the k-anonymity set shrinks [21]):

$$P_{ident} = \frac{1}{\text{number of profiles} - \text{profiles sorted out}}$$

As a result the user is uniquely identified with 100% accuracy if all profiles except the one of the user can be sorted out.

6.2. Bringing the User Back Into Control

In [13] we propose the integration of security and privacy information into "Context Descriptive Security Models" of the users' surroundings (see Fig. 7 for an example). These security models present the users' surroundings in a map like interface. In a supermarket, for example, the shelves are shown, together with relevant sensors. This will help the user to realize where the different parts of the information system are placed and how their interplay is to be evaluated.

It is obvious that it is undesirable to walk around the ambient environments and continually look at the security model presented on the display of some mobile device. We rather see this as a mechanism to gain the users' trust. They are in a position to visualize the system, there is no magic or hidden functionality. Once the users believe this, they do no longer need to ensure that this is really true. This effect is similar to the trust in open source software: the *option* to look through the code is sufficient to gain the users' trust. They hope that someone will look through the code and verify its trustworthiness.

What are the sources of information for this visualization interface? We see several possible sources of information for this type of visualization. The selection of the most adequate ones is still an open research question.

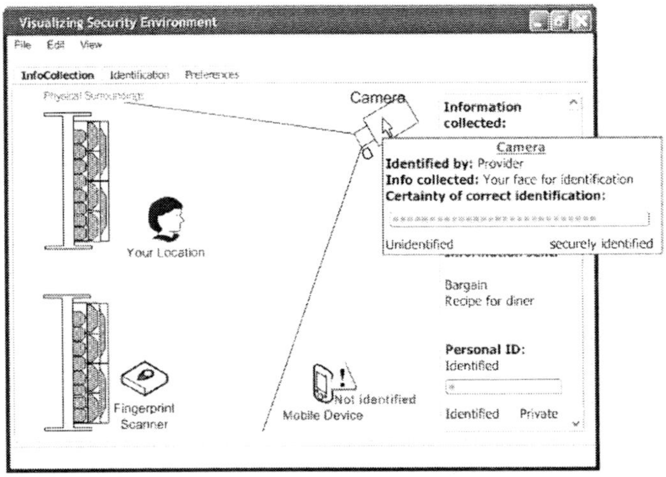

Figure 7. Mockup interface of the Context Descriptive Security Model.

According to the principles of social navigation [8], we integrate information from different sources into the map like representation of the users' environment. Social navigation in this context means, that decisions of other users are evaluated according to their outcome (with respect to the users' policy). This allows for the "visualization" of different paths and their effect on the map. In the real world this is like a worn path on a field of grass that leads us to the points of interest. Additionally, information gained from social navigation mechanisms can be used to calculate security measures like the anonymeter presented above. These are presented to the users in order to give them a view on the system's current state.

The foundations of secure logging mechanisms are the most basic building block. It will provide two possibilities to bring the user back into control again: On the one hand, the integration and evaluation of the interactions of other users with the system provide a great source of information. This evaluation is only possible if there exists enough reliable and trustworthy information about these interactions. On the other hand, the users have the possibility to reclaim their rights after they have been violated, if they are in a position to realize violations. To realize the infringement of users' rights and to reliably prove this infringement information written to the log file is essential.

Compared to the current state of the art in usability research, the approach presented in this paper provides several advantages. There exist interfaces that are signaling the current state of the privacy mechanisms [6]. While these work well for rather well defined interaction models, like browsing the web, the integration of these interfaces into the context is essential. Considering the different approaches for engineering usable and secure systems [9, 10, 11], it is important to realize that these well defined processes of developing, testing and deploying systems are no longer feasible for the steadily growing and evolving systems of ambient intelligence. The special properties of usable security as they are for example presented in [24], make it inevitable that dedicated user interfaces for security and privacy are carefully investigated and designed.

7. Conclusion

In this paper we have presented the scenario of the supermarket of the future. A plethora of sensors and computer systems enhance the shopping experience of the customers. We argue that it is inevitable to cope with the privacy, security and usability issues of these systems, because the pervasive nature of these systems does not allow customers to avoid these technologies. To this end, we present mechanisms to reliably and securely log the systems events and hence prove the systems' behavior. This brings the users into a position where they are able to realize possible infringements and they can actively control their privacy and security requirements. As an interface between the users and the system we propose "Context Descriptive Security Models". These allow for the intuitive presentation of measures of relevant security properties, such as anonymity or strength of confidentiality. They, furthermore, allow for the integration of information collected from other users and their evaluation of the outcome. This social navigation will help the users to find a configuration of their systems that really fits their needs.

References

[1] Accorsi, R.: On the relationship of privacy and secure remote logging in dynamic systems. In: Fischer-Hübner, S.; Rannenberg, K.; Yngström, L.; Lindskog, S.; editors, *Proceedings of the 21st IFIP TC-11 International Security Conference: Security and Privacy in Dynamic Environments,* International Federation for Information Processing. Springer-Verlag, 2006, 329–339.

[2] Accorsi, R.; Hohl, A.: Delegating secure logging in pervasive computing systems. In: Clark, J.; Paige, R.; Pollack, F.; Brooke, P.; editors, *Proceedings of the 3rd International Conference on Security in Pervasive Computing, Lecture Notes in Computer Science* **3934**. Springer Verlag, 2006, 58–72.

[3] Alahuhta, P.; De Hert, P.; Delaitre, S.; Friedewald, M.; Gutwirth, S.; Lindner, R.; Maghiros, I.; Moscibroda, A.; Punie, Y.; Schreurs, W.; Verlinden, M.; Vildjiounaite, E.; Wright, D.: *Dark scenarios on ambient intelligence: Highlighting risks and vulnerabilities.* Technical report, SWAMI Deliverable D2 A report of the SWAMI consortium to the European Commission under contract 006507, November 2005.

[4] Bellare, M.; Yee, B.: *Forward integrity for secure audit logs.* Technical report, University of California at San Diego, Dept. of Computer Science & Engineering, 1997.

[5] Blache, F.; Chraiet, N.; Daroux, O.; Evennou, F.; Flury, T.; Privat, G.; Viboud, J.-P.: Position-Based Interaction for Indoor Ambient Intelligence Environments, *Lecture Notes in Computer Science* **2875**. Springer, 2003, 192–207.

[6] Cranor, L.-F.; Guduru, P.; Arjula, M.: User interfaces for privacy agents. *ACM Transactions on Computer-Human Interaction*, 2006, 13(2), 135–178.

[7] Creese, S.; Goldsmith, M.; Harrison, R.; Roscoe, B.; Whittaker, P.; Zakiuddin, I.: Exploiting empirical engagement in authentication protocol design. In: Hutter, D.; Ull-

mann, M., editors, *Proceedings of the 2nd International Conference Security in Pervasive Computing, Lecture Notes in Computer Science* **3450**. Springer-Verlag, 2005, 119–133.

[8] DiGioia, P.; Dourish, P.: Social navigation as a model for usable security. In: SOUPS '05: *Proceedings of the 2005 symposium on Usable privacy and security,* 2005, 101–108.

[9] Dix, A. J.; Finlay, J. E.; Abowd, G. D.; Beale, R.: *Human Computer Interaction.* Prentice Hall, 2nd edition, 1998.

[10] Dourish, P.; Redmiles, D.: An approach to usable security based on event monitoring and visualization. In: *Proceedings of the 2002 Workshop on New Security Paradigms.* ACM Press, 2002, 75–81.

[11] Flechais, I.; Sasse, M. A.: Developing secure and usable software. In: *Workshop on Human-Computer Interaction and Security Systems. ACM,* April 2003.

[12] Friedewald, M.; Costa, O.D.; Punie, Y.; Alahuhta, P.; Heinonen, S.: Perspectives of ambient intelligence in the home environment. *Telematics and Informatics*, August 2005, 22(3).

[13] Höhn, S.: Bringing the user back into control. In: S. Fischer-Hübner, S. Furnell, and C. Lambrinoudakis, editors, Trust, Privacy and Security in Digital Business, *Lecture Notes in Computer Science* **4083**, Springer, 2006, 114–122.

[14] Itkis, G.: Cryptographic tamper evidence. In: *Proceedings of the Conference on Computer and Communication Security*, ACM Press, 2003, 355–364.

[15] Kaiser, J.; Reichenbach, M.: Evaluating security tools towards usable security. In: *Proceedings of the 17th IFIP World Computer Congress (WCC2002)*, August 2002.

[16] Lüke, F.: *Jobgefährdend.* Die Zeit, 2006.

[17] Nehmer, J.; Becker, M.; Karshmer, A.; Lamm, R.: Living assistance systems: an ambient intelligence approach. In: ICSE '06: *Proceeding of the 28th international conference on Software engineering*. ACM Press, 2006, 43–50.

[18] Norman, D. A.: *The psychology of everyday things.* Harper Collins Poublishers, 1988.

[19] Sackmann, S.; Strüker, J.; Accorsi, R.: Personalization in privacy-aware highly dynamic systems. *Communications of the ACM,* September 2006, 49(9), 32–38.

[20] Schneier, B.; Kelsey, J.: Security audit logs to support computer forensics. *ACM Transactions on Information and System Security,* May 1999, 2(2), 159–176.

[21] Sweeney, L.: k-anonymity: A model for protecting privacy. *International Journal on Uncertainty, Fuzziness and Knowledge-based Systems,* May 2002, 10(5).

[22] Valcourt, E.; Robert, J.-M.; Beaulieu, F.: Investiganting mobile payment: supporting technologies, methods, and use.In: *Proceedings of the IEEE International Conference on Wireless And Mobile Computing, Networking And Communications,* 2005, 29–36.

[23] Weiser, M.: The world is not a desktop. *Interactions*, 1994, 1(1), 7–8.

[24] Whitten, A.: Tygar, J. D.: Why Johnny can't encrypt: A usability evaluation of PGP 5.0. In: *8th USENIX Security Symposium*, 1999.

[25] Wolfram, G.: *The future of retail: A visionary outlook.* Technical report, METRO Group Future Store Initiative, November 2005.

In: Long-Term and Dynamical Aspects of Information... ISBN 1-60021-912-8
Editors: Andreas U. Schmidt et al, pp. 19-27 © 2007 Nova Science Publishers, Inc.

Chapter 2

ANALYSIS OF PRIVACY POLICIES FOR DYNAMIC SYSTEMS

Martin Kähmer*
Institute of Computer Science and Social Studies, Dept. of Telematics
University of Freiburg, Germany

Abstract

Advances in three aspects of computing, namely reachability, pervasiveness, and autonomy, lead to highly dynamic systems, whose components increasingly manage themselves. However, their security and trustworthiness must always be evaluated in the users' context and in tight association with his personal goals and current preferences, especially if their privacy is endangered. Therefore, this paper proposes an extension to the policy language EPAL by defining an operator to compare two policies and to determine the difference in them. This policy operator enables a user to check for additional permissions that would come along with the acceptance of a new or modified policy and can assist him in estimating the risk he puts his security or privacy in before adopting it.

1. Introduction

The growing potential to combine devices with different capabilities and purposes has led to the development of densely networked, decentralized systems. In these systems, components are becoming increasingly autonomous; they avidly collect data in various forms and adapt to their environment as well as to the users' requirements. Their changing and possibly conflicting demands lead to a continuous negotiation of requirements and ad hoc relationships. These features lay the technical foundation for highly dynamic systems and, thus, for a plethora of new applications building up an infrastructure of services [11].

The retail industry is one of the few areas that has recognized the imminent benefits of the use of dynamic systems and further developed the necessary technology for this. This readiness first appeared through the Internet, whereby stores became electronically available to everyone at any time. An improvement of logistical processes subsequently resulted

*E-mail address: kaehmer@iig.uni-freiburg.de

through the use of RFID tags and the respective infrastructure. Now, the retail industry faces a new development, namely the future store, which can be described as a highly dynamic system. When utilized for delivering tailored services to improve customer relationship dynamic systems do not just allow for the delivery of universal services, such as product finders or information searches. By taking the customers' context into account, such as the products already picked or the current position of the cart, or considering name, age or purchasing history stored on customers' PDAs, the supermarket can offer individualized or even personalized services which adapt to the current needs of the customers [16].

The first attempts to realize such stores are already under way. In Germany, the METRO-Group is developing the "Future Store". In this store, shopping trolleys are fitted with personal shopping assistants, i.e. computers with barcode readers and WLAN connection as interface to the store's information system. Once identified by means of customer cards, personalized services can be offered via the shopping assistants – for instance recommender systems and advertising.

However, this future also involves the collection and use of personal-related data. Sensors and cameras mounted around the store record for instance the movement and behavior of a customer, not only to protect the store against theft, but to also let the information gained in this way – e.g. a customer stands apparently undecided at the wine section – flow into a direct advertising campaign on a nearby display panel.

As such systems are continuously becoming an essential feature of everyday life, the consideration of their security and privacy management is highly important. The acceptance of these systems is greatly affected by the capability of the user to keep control over the usage of his personal resources and to observe the systems' adherence to his preferences.

2. Policies: User Still in the Loop

The ability of autonomous components to operate independently without constant human supervision and their adaptation to current context by automated reconfiguration seem to push the user out of the control loop. For their management, the interaction between user and system appears to be dispensable.

Yet, it is important to realize that the evaluation of security and privacy risks cannot be an issue for those expert security administrators alone, who set up some default guards constraining the system's behavior. Although the border of automation will be moved in such environments, the user cannot be completely released from the responsibility of formulating his goals and preferences as a personal security policy. Trustworthiness and risks cannot exclusively be assessed by central administrators, especially when the user's privacy is endangered. Hence, users must have the opportunity to determine for themselves the extent of his data collection and the way of use [18].

In the retail, a range of stores do meanwhile present general behavior guidelines for their establishments, but the use of a given set of rules contradicts the often very nuanced and unconscious control of personal data in everyday life [2]. In order not to be at the mercy of such one-sided promises, the customer must actively control the behavior of the dynamic system. Carrying his own individual privacy policy stored on a suitable communication device, such as a customer card or a personal PDA, in addition to personal and personal-related data, the customer can actively regulate the collection and further use of his data.

By means of his PDA on site or applications on his personal computer at home, he can formulate and adapt regulations like the following according to his individual requirements and wishes:

- "Allow the analysis of the food shopping list for customer-related advertising and recipe suggestions, but delete any data collected after purchasing has been completed."
- "Only allow video surveillance when recorded data is directly anonymized by making the face unrecognizable or inform customers."
- "Forbid data evaluation if video surveillance is active."

Taking stock, the user needs to transfer his personal policy to the system. However, his policy will often not be the only clincher for the system's security decisions. If there is a set of global security rules in place guarding the environment, the user will not be able to overrule it. The system will more likely incorporate the user's policy and subordinate its rules to the global ones. Similarly, the resulting policy shared by the system components may be a conjunction of multiple user policies, if the system is used by more than one user at the same time. In consequence, the high dynamics of such systems forbid this shared policy to be permanently determined and, therefore, to be analyzed for the components' security behavior in advance [1]. A user cannot predict the future system composition. Its parameters may change so quickly that a determined and subsequently analyzed policy becomes useless after minutes or even seconds.

More importantly, users are often unable to predict even their own future behavior. As surveys have shown, users take a very pragmatic approach to security. Willing to make security and privacy trade-offs according to the current situation or to the current communication partners, they adapt their own policies or spontaneously commit their resources and personal data to a foreign one to reflect the dynamics of such systems [16, 7].

Since users who have a method to observe and preview the result of their interaction make significantly less security-related operating errors [10], they need appropriate support for adapting the often complex policy sets and for evaluating the impact of the changes made. A step in this direction is the definition of the policy difference presented in the following chapter. This difference operator allows the user for the reduction of a modified or updated policy to only those parts that are relevant to him for his further security negotiations and decisions.

3. Analysis of Policies

A language for privacy policies is subject to certain demands on its strength of expression, i.e. which situations can be described. Taking up the store example, a policy must therefore express which service or user (a particular cashier) may carry out an action (reading) from a data set (customer data), for what purpose (billing) in what situation (on duty). As the use of personal and personal-related data is concerned, it is important to not just be able to

[1] Technical or legal policy enforcement is not within the scope of this paper; see [8] for further information.

regulate the data access, but also to impose duties, i.e. obligations, on data collectors. Only through the extension of access control to usage control the user can actively regulate the subsequent use of his data and demand, for example, the erasure of the collected data after purchasing has been completed [13].

Usage control policies integrate the diverse security policy concepts into a unified policy model [12]. With this, the user is not just able to declare who may access his resources, i.e. his own system components and his personal data. By means of obligations, he can also control their usage once they have been handed over to a foreign security domain. Combined with abstracting elements and categorizing them into hierarchies found in many access control languages, such policies are ideal for the intuitive specification of users' requirements in dynamic systems.

3.1. Overview of EPAL

Featuring both, the specification of obligations and the logical aggregation of resources by forming hierarchies, IBM's policy language EPAL [3] respectively its underlying algebraic analog E-P3P [5] build the basis for our definition of policy difference, though the underlying concept can be easily transferred to other usage control languages. For better understanding of the next section, some properties of the abstract syntax and semantic of EPAL are presented in the following, but definitively without the claim of completeness.

Hierarchies for users U, data D, purposes P and actions A build the vocabulary Voc together with conditions $C(Var)$, and obligations O. Similar to hierarchies, an obligation model defines an imply-relation on the power set of all obligations indicating that fulfilling a set of obligations implies fulfilling all of its subsets. Given a vocabulary, a policy $\mathscr{P} = (Voc, R, gc, dr, do)$ consists of a global enabler $gc \in C(Var)$, the default dr of ruling $r \in \{+, \circ, -\}$ according to "allow", "do not care", and "deny", a set of default obligations $d\bar{o}$, and an ordered list of rules $R \in \{U \times D \times P \times A \times C(Var) \times r \times \wp(O)\}$. A rule matches a request $q \in \{U \times D \times P \times A\}$, if its conditions are satisfied by the given assignment χ of environment variables, and all elements of the request are equal to or children of their rule counterparts. The first matching rule in $\mathscr{P}$ gives the result $eval_{\mathscr{P}}(q, \chi) = (r, \bar{o})$ consisting of a ruling and a set of obligations. For a complete specification, see [5].

3.2. Policy Difference

Whenever a user joins a dynamic system he is faced with the question of whether to accept its policy or to decline it and, perhaps, enter a further round of policy negotiation. Similarly, a user needs to follow the changes in a policy over time. If he has closely studied the policy of a service at an earlier point of time and found it acceptable, it is of interest to him as to whether the policy has meanwhile changed, e.g. by merging it with a another personal one, and whether it now makes the service more concessions. For this decision, both old and new policy versions are to be compared with each other.

Therefore, a user must have functions at his disposal with which he can analyze and compare the policies with regard to their assertions. For the comparison of two policies, EPAL defines the equivalence of two policies by means of the results of their evaluations: Two policies are equivalent if they provide the same decisions for all possible queries and

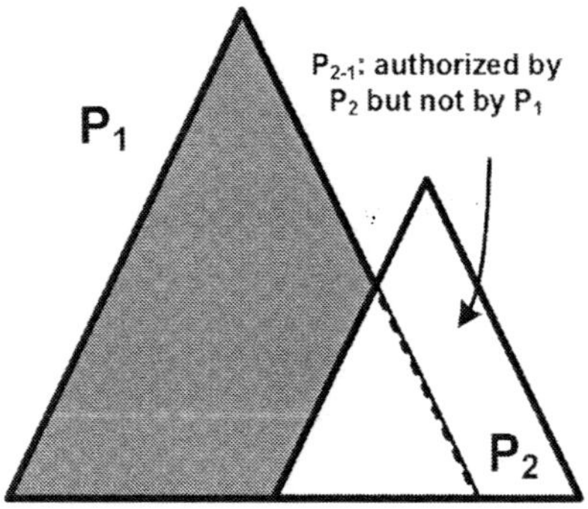

(a) $\mathscr{P}_2$ grants further actions, hence the (white) difference $\mathscr{P}_{2-1}$ is not empty.

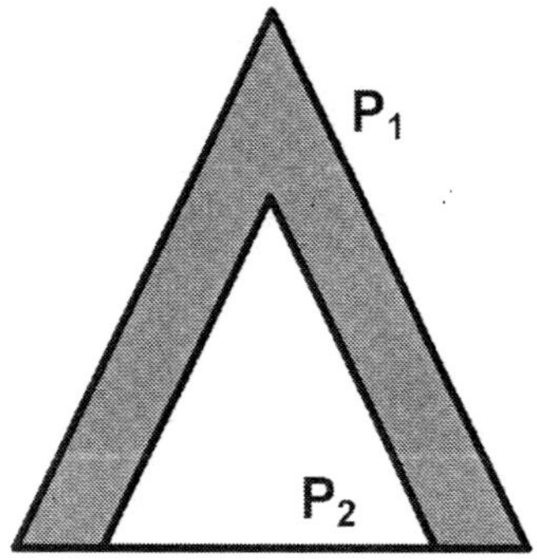

(b) $\mathscr{P}_{2-1}$ is empty, i.e. $\mathscr{P}_2$ is completely covered by $\mathscr{P}_1$ and weakly refines it.

Figure 1. Difference and refinement of two policies.

allocations of the environment parameter. The refinement test is defined in a similar way. With it, it can be examined whether a policy is more restrictive or more specific than another, therefore fulfills this other policy on all points. If the new version is equivalent to or even more restrictive than the old one, an automated process can be used to indicate acceptance.

For dynamic systems, however, these two tests are not sufficient, for the user is confronted right here with many services and their respective individual policies and it is not to be assumed that all the same or, at the most, more restrictive decisions than those of the user are made. How should the user behave if the policy of a service does not correspond to his own, i.e. it is not equivalent or more restrictive? The user must intervene at this point and analyze both policies more extensively. He must ascertain what rights in what situations can additionally be demanded by him so that the service can be executed. For example, the vendor may not be only allowed to access the user's food shopping list but all other lists, such as the one for the chemist round the corner, too. With the knowledge of this difference, the user can now decide whether he wants to dispense with the service of the vendor or disclose the complete shopping list.

To prevent him from being overwhelmed by complex policy sets, the difference in the new and old policy reveals to him only those permissions granted by the former, but not by the latter, i.e. a reduction to only those permissions relevant for consideration at this time. A non-empty difference indicates the extra permission that is not covered by the regulations $\mathscr{P}_1$ he has set for the treatment of his data, but that the systems demands for the usage and tailoring of its services in policy $\mathscr{P}_2$ (cf. Fig. 1(a)). Otherwise, the user can verify by an empty difference that the policy $\mathscr{P}_2$ is more restrictive and, thus, safe to use (cf. Fig. 1(b)).

Fig. 2 shows an example reduced to the consideration of the data hierarchy: $\mathscr{P}_1$ allows some actions on data D_1, whereas $\mathscr{P}_2$ allows the same actions on D_2 and D_4. Computing the difference, written $\mathscr{P}_{2-1}$, reveals that $\mathscr{P}_2$ extends the permission to D_2 and its children D_5 and D_6, but not to D_4, as it is already covered by the permission on D_1 by $\mathscr{P}_1$. From the user's point of view, the permission on D_1 and D_3 taken by $\mathscr{P}_2$ (i.e. $\mathscr{P}_{1-2}$) are not of primary interest, because no possibly unintended permission is inserted.

Assuming a single vocabulary for both policies the notion of policy difference is defined in style of [4] as follows:

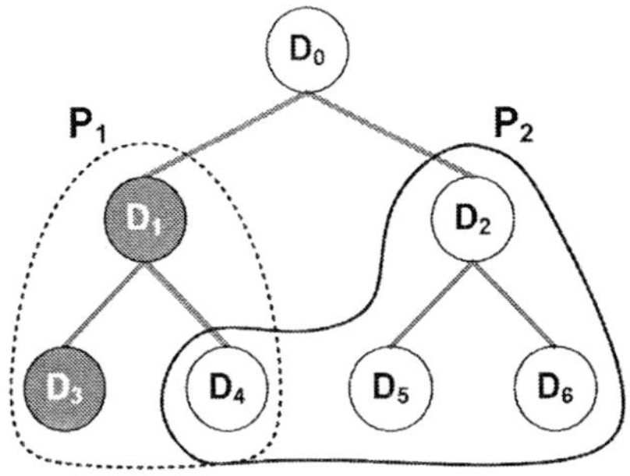

Figure 2. $\mathscr{P}_1$ allows some actions on D_1 and, thereby, D_3 and D_4, $\mathscr{P}_2$ on D_4, D_2, and both its children.

Definition (Policy Difference). *Let two policies $\mathscr{P}_1$ and $\mathscr{P}_2$ be given with $\mathscr{P}_i = (Voc, R_i, gc_i, dr_i, d\overline{o}_i)$ for $i = 1,2$ and equal vocabulary. Then, the rule set of the difference $\mathscr{P}_{2-1}$ gives the same query results as $\mathscr{P}_2$, but it is only enabled for those queries q and assignments χ of condition variables, for which one of the following statements holds, where $(r_i, \overline{o}_i) = eval_{\mathscr{P}_i}(q, \chi)$ for $i = 1,2$:*

- $gc_1(\chi) = false \quad and \quad (r_2 \in \{+, \circ\} \quad or \quad \overline{o}_2 \neq \emptyset)$
- $\overline{o}_1$ *does not imply* $\overline{o}_2$
- $r_1 = \circ$ *and* $r_2 = +$
- $r_1 = -$ *and* $r_2 \in \{+, \circ\}$

A first implementation can be made using a brute force approach. One simply evaluates both input policies for any possible request q and any possible assignment χ and compares their rulings and obligation sets. A rule $(q, C_2, r_2, \overline{o}_2)$ generated from a given q and χ is part of $\mathscr{P}_{2-1}$, if policy $\mathscr{P}_1$ is globally disabled and, therefore, not active in context χ, but $\mathscr{P}_2$ does not prohibit the requested action or commits the requester to future actions. For all other q and χ, the generated rules are part of the difference, unless $\overline{o}_2$ does not contain any new obligation and is a subset of $\overline{o}_1$. In this case, it is up to the comparison of the rulings only. If r_1 is dominated by r_2, i.e. r_2 is more permissive, the corresponding rule is part of $\mathscr{P}_{2-1}$, too.

As this brute force approach tests all valid combinations of requests and condition variables, their amount grows exponentially with the number of vocabulary variables. Thus, a more efficient algorithm is desired, e.g. by grouping requests with the same matching rules or neglecting rules that will always be hidden by matching rules with higher priority, as Backes et al. propose for their efficient algorithm to test on policy refinement [4]. Exploiting the semantic of the policy language or rather its evaluation function to reduce the difference problem to only the relevant combinations is part of future work.

4. Related Work

Currently, the comparison of such policies is made by a refinement check [5], which enables the automated verification that one policy also fulfills the other one or, in case of weak refinement, that the former is less permissive. For this test, an efficient algorithm is given in [4], but in case of verification failure, the user gets no hint of its cause. Furthermore, weak refinement can be mapped to the evaluation of the policy difference: if $\mathscr{P}_{2-1}$ is empty, then $\mathscr{P}_2$ is more restrictive and weakly refines $\mathscr{P}_1$ (cf. Fig. 1(b)).

For the limited name-space of P3P, the Privacy Bird [7] implements a refinement test. Preferences are specified as a selection from a predefined set of high-level rules the bird tests separately against the site's policy. If one of them is not refined, its description tag is shown to the user. Using policy difference, the conflicting parts are indicated to the user without any clever name-space separation ex ante.

There is ongoing work on the evaluation of policy changes and their compliance with meta rules, such as "separation of duty" [17], but research is focused on RBAC and corresponding algebras only.

Finally, there are approaches using model checking with dynamic logic [14] or temporal logic [6]. While these approaches allow for an elegant representation of policy changes and for checking policy compliance of the system, a formal model of the system is needed, which will be difficult, if not impossible to acquire in a highly dynamic system.

5. Conclusion

In this paper, we put forward the users' point of view on policies in dynamic systems. We argue that despite their autonomous behavior the components still need supervision of the users. While users adapt their usage control policies over time, they need support for observing the possible impact of the changes. For this, we define an operator to determine the difference in two policies, which reveals exactly those permissions additionally granted by the one and not the other. This definition is just the first step in the direction of usable policy management, and there is considerable work to be done. Subsequently, an efficient algorithm for computing the difference is to be developed. Finally, it will be transferred to syntax and semantic of the NAPS privacy policy framework [15], which enhances EPAL's underlying algebra by closure under composition and conjunction, both of them being essential policy operations in dynamic systems.

This approach is part of a project in which the issue of privacy protection in highly dynamic systems is examined. In order to make the visualization of the policy differences comprehensible to user and to assist him in his further security decisions, this information must be accordingly processed. An indicator presents a possible form of realization which, in the event of privacy violation, also shows the actual differences [9]. Furthermore, such policies are also essential for the preparation of so-called "privacy evidence" that contains information about data already collected and its use [16]. In interaction with secure logging protocols [1], policies specify which events must be recorded in order to ensure the integrity of the evidence. A simulation is currently being worked upon where this interaction is to be examined in practice.

References

[1] Accorsi, R.: On the relationship of privacy and secure remote logging in dynamic systems. In: Fischer-Hübner, S.; Rannenberg, K.; Yngström, L.; Lindskog, S.; editors, *Proceedings of the 21st IFIP TC-11 International Security Conference: Security and Privacy in Dynamic Environments*, Springer-Verlag, 2006, pp. 329-338.

[2] Ackerman, M. S.; Mainwaring, S. D.: *Security and Usability, chapter Privacy Issues and Human-Computer Interaction.* O'Reilly Media, 2005, pp. 381-399.

[3] Ashley, P.; Hada, S.; Karjoth, G.; Powers, C.; Schunter, M.: *Enterprise privacy authorization language (EPAL 1.2).* Submission to W3C, 2003.

[4] Backes, M.; Karjoth, G.; Bagga, W.; Schunter, M.: Efficient comparison of enterprise privacy policies. In: *Proceedings of 2004 ACM Symposium on Applied Computing*, ACM Press, 2004, pp. 375-382.

[5] Backes, M.; Pfitzmann, B.; Schunter, M.: A toolkit for managing enterprise privacy policies. In: *Proceedings of 8th European Symposium On Research In Computer Security (ESORICS)*, LNCS 2808, Springer-Verlag, 2003, pp. 162-180.

[6] Barth, A.; Datta, A.; Mitchell, J. C.; Nissenbaum, H.: Privacy and contextual integrity: framework and applications. In: *Proceedings of 2006 IEEE Symposium on Security and Privacy*, IEEE Computer Society, 2006, pp. 184-198.

[7] Cranor, L. F.; Guduru, P.; Arjula, M.: User interfaces for privacy agents. *ACM Transactions on Computer-Human Interaction,* 2006, 13(2), 135-178.

[8] Hilty, M.; Basin, D.; Pretschner, A.: On obligations. In: *Proceedings of 10th European Symposium On Research In Computer Security (ESORICS)*, LNCS 3679, Springer-Verlag, 2005, pp. 98-117.

[9] Höhn, S.: Bringing users back into control: A new paradigm for usability in highly dynamic systems. In: S. Fischer-Hübner, S. Furnell, and C. Lambrinoudakis, editors, *Proceedings of the 3rd International Conference on Trust, Privacy and Digital Business*, LNCS 4083, Springer-Verlag, 2006, pp. 114-122.

[10] Kaiser, J.: *Besteht eine Beziehung zwischen Nutzbarkeit und Sicherheit?* PIK "Sicherheit", 2003, 26(1), 48-51.

[11] Müller, G.; Accorsi, R.; Höhn, S.; Kähmer, M.; Strasser, M.: *Die Informatisierung des Alltags. Leben in smarten Umgebungen.* Chapter Sicherheit im Ubiquitous Computing: Schutz durch Gebote? Springer-Verlag, to appear 2007.

[12] Park, J.; Sandhu, R.: The UCONABC usage control model. *ACM Transactions on Information and System Security (TISSEC)*, 2004, 7(1), 128-174.

[13] Pretschner, A.; Hilty, M.; Basin, D.: Distributed usage control. *Communications of the ACM,* 2006, 49(9), 39-44.

[14] Pucella, R.; Weissman, V.: Reasoning about dynamic policies. In: *Proceedings of 7th International Conference on Foundations of Software Science and Computation Structures (FOSSACS'04)*, 2004.

[15] Raub, D.; Steinwandt, R.: An algebra for enterprise privacy policies closed under composition and conjunction. In: Müller, G.; editor, *Proceedings of International Conference on Emerging Trends in Information and Communication Security (ETRICS)*, LNCS 3995, 2006, pp. 132-146.

[16] Sackmann, S.; Strücker, J.; Accorsi, R.: Personalization in privacy-aware highly dynamic systems. *Communications of the ACM,* 2006, 49(9), 32-38.

[17] Schaad, A.; Moffett, J. D.: A lightweight approach to specification and analysis of role-based access control extensions. In: *Proceedings 7th ACM Symposium on Access Control Models and Technologies (SACMAT)*. ACM Press, 2002, 13-22.

[18] Tichy, G.; Peissl, W.: *Grundrechte in der Informationsgesellschaft.* Volume ITA-01-01 of manuscript, chapter Beeinträchtigung der Privatsphäre in der Informationsgesellschaft. Institut für Technikfolgen-Abschätzung, Österreichische Akademie der Wissenschaften, 2001, 22-48.

In: Long-Term and Dynamical Aspects of Information... ISBN 1-60021-912-8
Editors: Andreas U. Schmidt et al, pp. 29-44 © 2007 Nova Science Publishers, Inc.

Chapter 3

Preserving Privacy in the Web by Using Information Flow Control

Dieter Hutter*
German Research Center for Artificial Intelligence
66123 Saarbrücken, Germany

Abstract

The vision of a landscape of heterogeneous web services deployed as encapsulated business software assets in the Internet is currently becoming a reality as part of the Semantic Web. When pro-active agents handle the context-aware discovery, acquisition, composition, and management of application services and data, ensuring the security of customers' data becomes a principle task. To dynamically compose its offered service, an agent has to process and spread confidential data to other web services demanding the required degree of security. In this paper we propose a methodology based on type-based information flow to control the privacy of data and their proliferation within a set of communicating web services.

1. Introduction

The proliferation of web services as self-contained web accessible programs and the idea of the Semantic Web of making information *computer-interpretable* enables the dynamic composition of complex services assembled from various individual services and typically distributed over the web. Following the paradigm of pervasive computing, pro-active agents are installed on mobile phones or PDAs operating on the web and handle the context-aware discovery of appropriate services and use AI-based plan generation techniques to dynamically compose the retrieved service to solve complex tasks. Functional composition of complex services is well understood and supported.

However, the introduction of web services in general and dynamic web service compositions (e.g. [3, 25]) in particular requires appropriate security facilities to guarantee the security requirements of all participants. On the one hand, web services have to be protected

*E-mail address: hutter@dfki.de

against misuse of their resources, and on the other hand, the user of web services require the privacy of their data. Standard approaches for secure execution of services such as those using REI [20] or Ponder [6] are based on the specification of access control policies to control the individual execution of services and thus focus on the first task. We propose an approach [11, 12] to ensure the privacy of user data by introducing a dynamic security check between plan generation and execution of the plan. Thereby we can guarantee that the execution of a synthesized plan will not distribute user's or newly generated data to a non-entitled web service. We make use of techniques developed in program language security in general and an adapted version of Volpano Smith's [26, 27] security type calculus in particular.

To protect the privacy of user related information, the data used in web services are always annotated with their security policy. Hence, data is always *classified* according to its confidentiality and web services require the corresponding *clearances* to deal with confidential data. Both classifications and clearances denote together the *sticky policy* of the data which is formulated by the help of so-called *security types*. There is a partial ordering $\leq$ on security types which allow us to compare them. The set of all security types together with $\leq$ forms a lattice. Similar to the approach presented by Bell and LaPadula [2], the idea is that a web service is only entitled to obtain a specific datum if its clearance is at least as high as the classification of the data.

However, in practice we would like to select the clearances of web services and also the classification of data with respect to individual categories or types of information. For instance, while we trust a doctor to keep our medical health status confidential we might have mixed feelings when providing the same person with a direct debit authorization for our bank account. Analogously, a given datum may allow us, for instance, to infer confidential information about the location of a person or confidential information about his bank account. Hence, both classifications of data and clearances of web services are described by a vector of security types. Each entry in the vector denotes the classification or clearance with respect to a particular category or roles (like, for instance, location or payment information).

2. An Example

Throughout this paper we will make use of a running example illustrating our approach for enforcing privacy in the semantic web. Suppose a user called Peter Miller is suffering from a cold and Aids, and wants to keep this fact private as far as it gets, that is no one whom he does not trust shall know about his disease, in particular that he is suffering from aids. Peter calls his personal medical agent (MED) via a mobile phone, requesting it to purchase appropriate medicine online at a relevant online pharmacy in the web. The required drug for treatment of his disease has to be a special one in the sense that it has to take both cold and Aids into account at the same time. As to the provision of web services, we assume that one or multiple web services are encapsulated and deliberately provided by so called service agents. As a consequence, the Peter's personal agent MED first has to discover the set of available online pharmacy service and payment service agents. It then selects the most relevant pharmacy agent to obtain information on the appropriate drug and its price, and then makes the corresponding payment online.

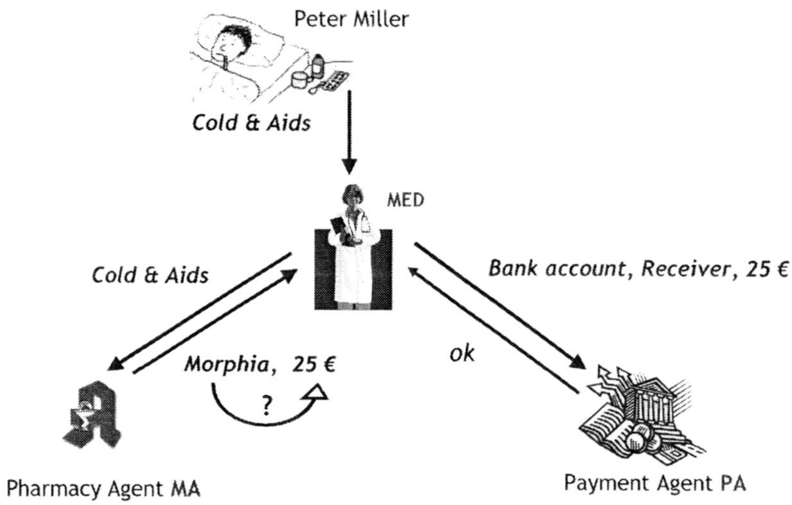

However, the resulting composed service to be jointly executed by the agents MA and PA (see Figure 1) has to be checked whether it complies with Peter's privacy concerns. For example, the pharmacy agent MA could disclose the type of Peter's disease to other agents. If the price to be announced to the payment agent PA depends on the drug and this drug discloses the disease, PA could indirectly deduce Peter's disease from the order for payment including the price and the identity of Peter Miller as purchaser. Two solutions to this problem are obvious. Either the price of the drug does not reveal the type of the drug (resulting in some kind of "flatrate" for medicine) or Peter has to trust the payment agent with respect to his health status.

In order to verify this appropriateness we use information flow analysis to verify that sensible information will flow only to authorized agents with appropriate clearance. We will explain this approach in more detail in subsequent sections and illustrate it with the help of this use case scenario.

3. Sticky Policies for Data

The basic idea of sticky policies [13, 24] is to attach security policies to individual data that define the further distribution of this data. Services receiving data have to obey their policies once they want to process or forward them. Compiling new information or manipulating existing data results in the inheritance and merge of the security policies of the involved data. In our example Peter Miller has to specify who is authorized to know that he suffers from aids. In order to do so, we adopt the idea of security levels as presented by Bell and LaPadula in the seventies. Thus security policies are formulated by assigning clearances of web services and classifications to data. A web service is allowed to access a datum or information only if its clearance is higher than or equal to the classification of the data.

The individual values of classifications and clearances are called security types. The set of all security types forms a lattice. Hence a security typeset is a partially ordered set $(L, \leq)$ such that for each two elements in L there is a least upper bound (join) and a greatest lower bound (meet). We additionally require that L has a least element $\bot_L$ which we can use for delegation issues. In our use case we use a simple lattice consisting of the security types $\perp$, L, and H with $\perp \leq L < H$. Peter Miller classifies the information that he suffers from Aids as H. Web services that he trusts will receive an H-clearance for this data while others will obtain the clearance L. Since Peter does not know all the web services that are available, all unknown web services are automatically considered as $\perp$. Distinguishing between $\perp$ and L allows us to dsistinguish between web services that are unknown to Peter and those that are distrusted by him. While delegation of rights may result in an upgrading of previously unknown web services, trusted web services must not upgrade web services Peter distrusts.

In the following we introduce web spaces as collections of web services.

Definition 1. *A* web space W *is a finite set of web services.* $\mathcal{W}$ *denotes the set of all web spaces.*

In order to specify clearances of web services we introduce mappings from web spaces to security types.

Definition 2. *Let* $W \in \mathcal{W}$ *and* $(L, \leq)$ *be a lattice denoting the security types. An* assignment σ *of* W *with respect to* $(L, \leq)$ *is a mapping from* W *to* L. $DOM(\sigma) = \{w \in W \mid \sigma(w) \neq \perp_L\}$ *is called the domain of* σ.

Initially each user or web service possesses individual personal data. For each personal datum an individual security policy is specified by fixing its classification and assigning clearances to the known web space with respect to this datum. In this way an information provider (typically the user) can define which web service she trusts with respect to this data (by assigning it a clearance higher or equal than the classification of the data).

In our example, we define the web space W simply as $W = \{MED, MA, PA\}$. Then an assignment of W formalizing Peter's assessment of who may know his disease would map MED and MA to H and PA to L. The information about his disease would be classified as H.

Processing data will result in new compiled information. Its resulting security policy has to combine the security policies of the data used to compile it. Once the pharmacy agent selects a drug for Peter that is specific to his disease then the information that Peter orders this drug inherits the classification of the knowledge of Peter's disease. The information that Peter needs this particular drug discloses also information about Peter's disease. In a similar way the price of the drug may reveal information about the drug itself. Knowing the prices of all available drugs and knowing the price of Peter's drug we can speculate which drugs Peter has ordered and based on this information we can also speculate about his disease. Beside Peter also the pharmacy has its privacy concerns. It may operate a specific rebate system for Peter and does not want to disclose this system to the public. Hence it may formulate its own policy of who may see the price of the drug. Then, to see the price a web service must be trusted by Peter and the pharmacy. In general, a computed information (like the price of the drug) can be forwarded to a web service if at least all the data used to compute this information would be accessable to this particular web service.

Therefore, data are annotated by multiple security policies that are inherited from the security policies of the different sources used to process this data. We combine the security policies by combining the underlying lattices in the following way:

Definition 3. *Let $(L_i, \leq_i)$ with $1 \leq i \leq n$ be lattices. Then, the* combined *lattice of a sequence $(L_1, \leq_1), \ldots, (L_n, \leq_n)$ is the partially ordered set $(\mathscr{L}, \leq)$ with $\mathscr{L} = L_1 \times \ldots \times L_n$ and $(l_1, \ldots, l_n) \leq (l'_1, \ldots, l'_n)$ iff $l_i \leq_i l'_i$ for all $1 \leq i \leq n$.*

It is obvious that a combined lattice is itself a lattice if we specify

$$lub((l_1, \ldots, l_n), (l'_1, \ldots, l'_n)) = (lub(l_1, l'_1), \ldots, lub(l_n, l'_n))$$
$$glb((l_1, \ldots, l_n), (l'_1, \ldots, l'_n)) = (glb(l_1, l'_1), \ldots, glb(l_n, l'_n)).$$

Another reason to use combined lattices to assess clearances and classifications is to simplify the specification of security policies in the first place. Typically, a user will use different (fixed) categories to classify his data. A data may contain details about our financial status or may reveal information about our health status. As a user we may want to assign a web service different clearances depending whether the data contains financial or health information or even both. Peter trusts his bank with respect to financial issues but he does not want to reveal his medical condition to it. Vice versa, Peter formulated a clearance for the pharmacy allowing it to access Peter's medical records but not his bank account balance. Using combined lattices a user can statically assign different clearances to web services with respect to the different categories. Later on he has only to combine the clearances of the affected categories and provide corresponding classifications for the data.

Using combined lattices we extend the notion of assignments to sequences of lattices in the following way.

Definition 4. *Let $\mathscr{L}$ be the combined lattice for a sequence $(L_1, \leq_1), \ldots, (L_n, \leq_n)$ of lattices and σ_i be assignments of W wrt. $(L_i, \leq_i)$ for all $1 \leq i \leq n$. Then, the sequence $\Sigma = (\sigma_1, \ldots, \sigma_n)$ is called the* assignment space *of W wrt. $\mathscr{L}$.*
Let $w \in W$ be a web service then $\Sigma(w) = (\sigma_1(w), \ldots, \sigma_n(w))$ is its clearance *wrt. Σ.*

Intuitively the assignment space comprises the different assessments of trust that various information providers have in a common web space with respect to various categories. Each lattice $(L_i, \leq_i)$ reflects the set of security levels given by an individual provider and σ_i assigns each web service its clearance with respect to this provider.

Definition 5. *Let $\mathscr{L}$ be a combined lattice for a sequence $(L_1, \leq_1), \ldots, (L_n, \leq_n)$ and σ be an assignment space of W wrt. $\mathscr{L}$. Furthermore, let $\tau = (\tau_1, \ldots \tau_n)$ be a sequence of security types with $\tau_i \in L_i$ for all $1 \leq i \leq n$. Then, (σ, τ) is called a* sticky policy *with respect to $\mathscr{L}$.*
A web service $w \in W$ satisfies a sticky policy (σ, τ) iff $\tau_i \leq_i \sigma_i(w)$ for all $1 \leq i \leq n$.

Suppose, that Peter has two assignments of W denoting his trust with respect to medical issues and with respect to financial aspects. Concerning medial issues he uses an assignment σ_{med} with $\sigma_{med}(MED) = H, \sigma_{med}(MA) = H, \sigma_{med}(PA) = L$ while σ_{fin} with $\sigma_{fin}(MED) = H, \sigma_{fin}(MA) = L, \sigma_{fin}(PA) = H$ specifies his assignment with respect to financial issues. His classification of the information about his disease is formalized as $\tau = (H, L)$. This means that it contains confidential information about his medical status but only public (or even none) information about his financial status. While MA satisfies the sticky policy $((\sigma_{med}, \sigma_{fin}), \tau)$, PA does not because it has no sufficent clearance to receive information classified as H wrt. health status.

The set of categories used to invoke a web service changes typically while processing the request. When a called web service provides new information, it may introduce a new category and classify the new data also with respect to this new category. In this way a subsequently called web service can formulate its security requirements for its provided data by defining also the clearances of all web services with respect to the new category.

4. Formalization of Web Services

Sticky policies restrict the way private data can be exchanged and used between web services. Intuitively a web service should only obtain information for which it possesses the appropriate clearance. Furthermore, it should not be able to infer private information (for which it does not possess a sufficient clearance) simply by observing the behaviour of other web services. In the following we analyse the information flow in web spaces and develop sufficent criteria for individual web services to guarantee that no information is disclosed to an unauthorized web service.

We start with the modelling of web services and their interaction following the ideas in [7] as a collection of automata and continue with the formalisation of the requirements to individual web services such that the overall system will not leak any private data to unauthorized web services.

We specify a web service as a non-deterministic (possibly infinite) automaton that communicates with other web services with the help of input- and output-events. The communication is asynchronous, i.e. each web service possesses an input buffer and at any time other web services can send messages to this input buffer by performing output-events regardless of the internal state of the targeted web service. Using input-events a web service can deliberately withdraw a message from its input buffer. It is crucial to notice that the sequence in which the web service can withdraw messages from its input buffer is independent of the sequence in which they were written to the buffer. In this sense each input buffer denotes a multiset rather than a queue of messages.

To simplify matters we assume a unique form of all messages. A message contains information about the sender and receiver of the message, a sequence of data, its (functional) specification, and a sequence of sticky policies. These sticky policies specify the access rights to the individual data and regulate who may get knowledge about the existence of the message itself. On the one hand the environment might run the web services in sandboxes that do not allow web service to monitor the messages of other web services. On the other hand using an open network as the Internet to communicate between web service might allow them to monitor at least partially the flow of messages between web services. Using an individual sticky policy for messages we are able to model both cases.

Definition 6. *Let* $W = \{w_1, \ldots w_m\}$ *be a web space. The message space* $\mathcal{M}$ *of* W *is a set of tuples of the form* $(w, w', (t_1, \ldots, t_n), \Phi, sp)$ *where*

- w *is the sender and* w' *the receiver of the message with* $w, w' \in W$,
- $t_1, \ldots t_n$ *are the data of the message,*
- Φ *is some service description, and*

- $sp = (sp_0, sp_1, \ldots, sp_n)$ *with* $sp_i = (\sigma_i, \tau_i)$ *for* $0 \leq i \leq n$ *is a family of sticky policies.*

$I_{w',\mathcal{M}}$ *is the subset of* $\mathcal{M}$ *with receiver* w' *while* $O_{w,\mathcal{M}}$ *is the subset of* $\mathcal{M}$ *with sender* w.

The message space of a web space is the set of all messages that can be communicated between web services. Φ denotes the description of the requested service. Since we are interested in the security of web spaces we will ignore functional aspects of web service and therefore will also ignore any inner structure of a service description Φ. This is also reflected in the fact that we will not distinguish between different internal events of individual web services but abstract from these (functional) details by specifying a web service w as a non-deterministic (but not necessarily finite) system having only one internal event ε_w. Its input- and output events are members of the message space of its corresponding web space.

Definition 7. *Let* $\mathcal{M}$ *be a message space for a web space* W. *Then a* web service specification *for* $w \in W$ *is a nondeterministic automaton* $(S_w, s_{o,w}, E_w, I_w, O_w, T_w)$ *with* $E_w = I_w \cup O_w \cup \{\varepsilon_w\}$, $I_w = I_{w,\mathcal{M}}$, $O_w = O_{w,\mathcal{M}}$ *and* $T_w \subseteq S_w \times E_w \times S_w$.
A set $\mathcal{WS} = \{(S_w, s_{o,w}, E_w, I_w, O_w, T_w) \mid w \in W\}$ *of web service specifications is a* web space specification *of* W *w.r.t.* $\mathcal{M}$.

Given specifications of each individual web service, we combine these specifications to obtain a specification of the overall web space. Such a specification is again a non-deterministic automaton. A state of this automaton is a tuple consisting of the states of each individual web service and the contents of the input buffers. Furthermore the state also includes the history of all messages that have been sent within the web space. Notice that this history is only used to formalize the security requirements for the overall web space in general and for the individual web services in particular but it is not needed once we want to implement such a system. We distinguish three different types of events: internal, input and output events. Since we use asynchronous message passing, each event will always change only the state of one web service and in case of export events it will also change the input buffer of another web service.

Definition 8. *Let* $W = \{w_1, \ldots w_m\}$ *be a web space and* $\mathcal{WS}$ *be a web space specification of* W *w.r.t.* $\mathcal{M}$. *A configuration of* $\mathcal{WS}$ *is a* $2m+1$*-tuple* $(Q_1, s_1, \ldots Q_m, s_m, M)$ *with* $Q_i \in I^*_{w_i}$ *being the input buffer of* w_i *and* $s_i \in S_{w_i}$ *being the actual state of* w_i. $M \in \mathcal{M}^*$ *is the message sequence recorded by a global observer.*

Given two configurations $c = (Q_1, s_1, \ldots Q_m, s_m, M)$ *and* $c' = (Q'_1, s'_1, \ldots Q'_m, s'_m, M')$ *then* $c \to c'$ *iff there is a* $1 \leq j \leq m$ *such that*

internal step: $(s_j, \varepsilon, s'_j) \in T_{w_j}$ *and* $s_i = s'_i$ *for all* $i \neq j$. $Q_i = Q'_i$ *for all* $1 \leq i \leq m$, *and* $M = M'$.

input step: $(s_j, \alpha, s'_j) \in T_{w_j}$ *for some* $\alpha \in I_{w_j}$ *and* $Q_j = \{\alpha\} \uplus Q'_j$, *and* $s_i = s'_i$ *and* $Q_i = Q'_i$ *for all* $i \neq j$ *and* $M' = M$.

output step: $(s_j, \alpha, s'_j) \in T_{w_j}$ *for some* $\alpha \in O_{w_j}$ *and there is an index* i *with* $\alpha \in I_{w_i}$, $Q'_{w_i} = Q_{w_i} \uplus \{\alpha\}$, $Q'_{w_k} = Q_{w_k}$ *for all* $k \neq i$ *and* $M' = M \circ \langle \alpha \rangle$.

Notice that the above specification of web spaces does not presume any particular protocol for web service communication or web service negotiations.

5. Security of Web Space Specifications

In principle there are two ways information may be disclosed to unauthorized web services. On the one hand information can simply be sent to an unauthorized web service either on purpose (for instance by some trojan horse) or unintentionally being not aware about the criticality of the information. Access control policies are typically used to control the dissemination of data. Obviously sticky policies formalize access control rules that must be enforced by individual web services.

On the other hand an unauthorized web service can monitor its environment and can sometimes observe messages being sent around. Knowing about the functiality of other web services, an observer may speculate about situations in which these services can only send such messages. This analyis may reveal information about the data that is processed by the web services. In information flow control techniques were developed to analyse these abilities of an observer and we make use of such techniques to guarantee that a web service does not disclose private information by exposing a specific behaviour depending on this information.

We start our analysis of web space security with discussing access control issues arising from attaching sticky policies to messages.

Whenever a web service w sends a message to another web service w' then obviously w' must have sufficient clearances to receive this data. That means that obviously w' must have the clearance to observe the message as a whole denoted by sp_0 of the sticky policy sp and it must also possess the appropriate clearances formalized in $sp_1,\ldots,sp_n$ to receive the individual data $t_1,\ldots,t_n$ of the message. If both conditions are satisfies then we say that the message is AC-secure.

Definition 9. *A message* $(w,w',(t_1,\ldots,t_n),\Phi,sp)$ *is* AC-secure *iff* w' *satisfies the sticky policies* $sp_0,sp_1,\ldots sp_n$.

We regard a web space specification as AC-secure if it is always guaranteed that all messages sent around between web services are AC-secure. That means, that no web service will disclose secrets simply by sending it directly to an unauthorized web service.

Definition 10. *A configuration* $(Q_1,s_1,\ldots Q_n,s_n,M)$ *is* AC-secure *iff each element in* M *is AC-secure. A specification of a web space* $W=\{w_1,\ldots w_n\}$ *is* AC-secure *if all its reachable configurations are AC-secure.*

Besides the problem that a web service may send private information to an unauthorized web service, there is also the threat that the unauthorized web service can *deduce* confidential messages by analysing the communication or behaviour of other web services, i.e. by analyzing the sequence of messages it can observe. We adopt techniques of information flow control to analyse this threat.

Information flow control (e.g. [16, 29, 21, 15]) relies on the idea of modeling confidentiality and privacy of data as restrictions on the flow of information between different domains of a system. Starting with the work of Goguen and Meseguer [8, 9], the restrictions on information flow for deterministic systems have been formalized as independence properties between actions and observations of domains: Alice's actions are confidential wrt. Charly if his observations are independent of her actions, i.e. if Alice changes her actions this does not cause different observations for Charly. In this case Alice is said to be non-interfering with Charly. For non-deterministic systems, the intuition works backwards: Alice is possibilistically non-interfering with Charly if the observations of Charly can be explained by several, different behaviors of Alice. Thus, Charly's observation does not reveal which actions Alice has chosen.

The next definition formalizes the typical *view* of a web service with respect to the communication inside a web space. As mentioned before, each message is labelled with a particular sticky policy sp_0 that defines the visibility of the message to other web services. Accordingly we define the visible subset of a sequence of messages with respect to a web service w:

Definition 11. *Let $w \in W$ be a web service and δ be a sequence of messages then the subsequence $\delta|_w$ of δ visible to w is defined by:*

$$\begin{array}{rll} \langle\rangle & if \delta = \langle\rangle, & \\ e \circ (\delta'|_w) & if \delta = (e \circ \delta') and \tau_0 \leq \sigma_0(w) & where\ sp_0 = (\sigma_0, \tau_0)\ in\ e, \\ \delta|_w & if \delta = (e \circ \delta') and \tau_0 \not\leq \sigma_0(w) & where\ sp_0 = (\sigma_0, \tau_0)\ in\ e. \end{array}$$

Even if a web service cannot observe a particular message, its content, i.e. the parameters $t_1, \ldots, t_n$, could be disclosed to the web service via other channels, for instance, by copying the parameters into a (visible) mail to the web service. The sticky policies $sp_1, \ldots, sp_n$ regulate the distribution of the information encoded into the parameters. Two messages are regarded as equal by an observer if he has access to the same parts of both messages and both messages also coincide in the accessible parts.

Definition 12. *Let $e = (w_1, w_2, (t_1, \ldots, t_n), \Phi, sp)$ and $e' = (w'_1, w'_2, (t'_1, \ldots, t'_n), \Phi', sp')$ be two messages. e and e' are equal for a web service w, $e \approx_w e'$ for short, iff $sp'_0 = sp_0$ and for all $1 \leq i \leq n$:*

- $\tau_i \leq \sigma_i(w)$ *or* $\tau'_i \leq \sigma'_i(w)$ *implies* $sp_i = sp'_i$ *and* $t_i = t'_i$.

Two sequences δ, δ' of messages are equal for a web service w, $\delta \approx_w \delta'$ for short, iff $\delta|_w$ and $\delta|_{w'}$ have the same length and are pairwise equal for w.

Two sequences of events are equal for a web service w if the subsequences containing only the messages of the sequences are equal for a web service w.

The next definition formalizes the conditions on an individual web service w to guarantee that there is no unauthorized information flow caused by its behavior visible to some observer w'. The underlying idea is that its visible behaviour is independent of any information that is confidential for w'. Hence, whenever w exposes some visible behaviour knowing secret information then it must be possible that w would show the same visible behaviour if it would know completely different secrets or no secrets at all (bisimulation). Consider w performs a visible step e then there has to be a sequence δ of steps which simulate the behaviour of executing e in the other situation (knowing different secrets). However, the execution of δ must not depend on the benevolence of other web services. For example, w must be able to execute δ without requiring that a particular secret message has arrived from some other web service.

Definition 13. *Let W be a web space and $ws = (S_w, s_{o,w}, E_w, I_w, O_w, T_w)$ be the nondeterministic automaton of some web service specification. ws is locally* IFC-secure *iff for all $\overline{w} \in W$ there is an equivalence relation $\approx_{\overline{w}}$ on S_w such that for all $s_1, s_2 \in S_w$ and all $e \in E_w$ the following holds:*

- *if $s_1 \approx_{\overline{w}} s_2$, $s_1 \longrightarrow_e s'_1$, $e \notin I_w$ and $\langle e\rangle|_{\overline{w}} = \langle e\rangle$ then there is an $\delta \in (E_w \setminus I_w)^*$ with $\langle e\rangle \approx_{\overline{w}} \delta$ such that $s_2 \longrightarrow_\delta s'_2$ and $s_2 \approx_{\overline{w}} s'_2$;*
- *if $s_1 \approx_{\overline{w}} s_2$, $s_1 \longrightarrow_e s'_1$ and $e \in I_w$ and $\langle e\rangle|_{\overline{w}} = \langle e\rangle$ then for all e' with $e \approx_{\overline{w}} e'$ there are $\delta, \delta' \in (E_w \setminus I_w)^*$ with $\langle e\rangle \approx_{\overline{w}} \delta \circ \langle e'\rangle \circ \delta'$ such that $s_2 \longrightarrow_{\delta\circ\langle e'\rangle\circ\delta'} s'_2$ and $s_2 \approx_{\overline{w}} s'_2$;*
- *if $s_1 \longrightarrow_e s_2$ and $\langle e\rangle|_{\overline{w}} = \langle\rangle$ then $s_1 \approx_{\overline{w}} s_2$, and*
- *if $s_1 \approx_{\overline{w}} s_2$, $s_1 \longrightarrow_{e_1} s'_1$, $s_2 \longrightarrow_{e_2} s'_2$ and $e_1 \approx_{\overline{w}} e_2$ then $s'_1 \approx_{\overline{w}} s'_2$*

The first two requirements postulate that any transistion of the automaton processing an event e can be simulated in an equivalent relation by some sequence δ of events such that w cannot distinguish e from δ. The third requirement requires that confidential events should not cause a change of the visible state and the last requirement ensures visibility of states is a congruence with respect to the visiblity of events.

These requirements to individual web services will allow us to guarantee that the overall web space is secure with respect to information flow control. In a first step we specify when two global states of a web space have the same visible behaviour for an indivdual web service w:

Definition 14. *Two configurations $c = (Q_1, s_1, \ldots Q_n, s_m, M)$ and $c' = (Q'_1, s'_1, \ldots Q'_n, s'_m, M')$ are equal for a web service w, $c \approx_w c'$ for short, iff $s_i \approx_w s'_i$ and $Q_i \approx_w Q'_i$ holds for all $1 \leq i \leq m$ as well as $M \approx_w M'$.*

Similar to individual web services, we consider a web space as secure with respect to information flow control if its visible behaviour for an individual web service w does not depend on any knowledge of information that is secret for w. Formally we define:

Definition 15. *A web space specification for W is IFC-secure with respect to a web service $w \in W$ iff for all reachable configurations c_1, c'_1 of $\mathscr{W}\mathscr{S}$ $c \approx_w c'$ and $c \longrightarrow_e \overline{c}$ there is a sequence of events δ such that $c' \longrightarrow_\delta \overline{c}'$ and $\overline{c} \approx_w \overline{c}'$ holds.*

A web space specification for W is IFC-secure iff it is secure with respect to all $w \in W$.

As the main theorem of this paper we can conclude that locally IFC-secure web services result in an IFC-secure web space.

Lemma 1. *Let $\mathscr{W}\mathscr{S}$ be a web space specification for a web space W. If each web service specification $ws \in \mathscr{W}\mathscr{S}$ is locally IFC-secure then $\mathscr{W}\mathscr{S}$ is IFC-secure.*

Proof 1. *Let $W = \{w_1, \ldots, w_m\}$ be a web space and $w \in W$. Let $c = (Q_1, s_1, \ldots Q_n, s_m, M)$ and $c' = (Q'_1, s'_1, \ldots Q'_m, s'_m, M')$ with $c \approx_w c'$. Thus $s_i \approx_w s'_i$ and $Q_i \approx_w Q'_i$ holds for all $1 \leq i \leq m$. Furthermore $M \approx_w M'$. Now suppose there is an event e with $c \longrightarrow_e \overline{c}$ then we have to construct an appropriate δ such that $c' \longrightarrow_\delta \overline{c}'$ and $\overline{c} \approx_w \overline{c}'$ holds.*

Case 1. Suppose e is an event of some web service w_i and $\langle e\rangle|_w = \langle\rangle$. Since w_i is locally secure we know that $\overline{s_i} \approx_w s_i$. Futhermore $\overline{s_j} = s_j$ for all $j \neq i$ such that $\overline{s_j} \approx_w s_j$ holds for *all* $1 \leq j \leq m$. *Since e is not visible to w changing the input buffer as an input or output event will still result in* $Q'_j \approx_w Q_j$ *for all* $1 \leq j \leq m$. *Hence,* $\overline{c} \approx_w c$ *and since* $c \approx_w c'$ *we know* $\overline{c}' = c' \approx_w \overline{c}$ *with* $\delta = \langle\rangle$ *and thus also* $\overline{M} \approx_w \overline{M}'$.

Case 2. Suppose e is an internal event of some web service w_i *and* $\langle e\rangle|_w = \langle e\rangle$. *Since* w_i *is locally IFC-secure and* $s'_i \approx_w s_i$ *there is a* $\delta \in (E_{w_i} \setminus I_{w_i})^*$ *such that* $s'_i \rightarrow \overline{s}'_i$ *with* $\overline{s}_i \approx_w \overline{s}'_i$. δ *can be processed locally at* w_i *except that it may write messages invisible to w to the input buffers of other web services. Let* $c' \rightarrow_\delta \overline{c}'$. *Since* $Q_j = \overline{Q}_j$, $Q_i \approx_w Q'_i$ *and* $Q'_j \approx_w \overline{Q}'_j$ *we know* $\overline{Q}_i \approx_w \overline{Q}'_i$ *for all* $1 \leq j \leq n$ *and as a consequence* $\overline{M} \approx_w \overline{M}'$. *Finally* $\overline{s_k}' = s'_k$ *for all* $k \neq i$ *and therefore* $\overline{s_k}' \approx_w \overline{s_k}$. *Thus, finally* $\overline{c}' \approx_w \overline{c}$ *holds.*

Case 3. Suppose e is an output event of some web service w_i *and* $\langle e\rangle|_w = \langle e\rangle$. *Again there is a* $\delta \in (E_{w_i} \setminus I_{w_i})^*$ *such that* $s'_i \rightarrow_\delta \overline{s}'_i$ *with* $\overline{s}_i \approx_w \overline{s}'_i$. *Since e is visible to w,* δ *has to contain an output event* e' *with* $e \approx_w e'$. *Thus, with k being the receiver of both output events e and* e' *it holds that* $\overline{Q}_k = Q_k \cup \{e\}$, $\overline{Q}'_k = Q'_k \cup \{e'\}$ *and therefore* $\overline{Q}_k \approx_w \overline{Q}'_k$. *Since* e' *is the only event visible to w also* $\overline{Q}_j \approx_w \overline{Q}'_j$ *holds also for all* $j \neq k$. *Analogously to the previous case,* w_i *can perform* δ *locally except that some output invisible to w may be written to some input buffers. Thus,* $\overline{Q}_j = Q_j \approx_w Q'_j \approx_w \overline{Q}'_j$ *for all* $j \neq k$ *and again* $\overline{s_j}' = s'_j$ *for all* $j \neq i$. *Since also* $\overline{M} = M \circ \{e\} \approx_w M' \circ \overline{\delta} = \overline{M}'$ *with* $\overline{\delta}$ *being the output events of* δ. *Thus, finally* $\overline{c}' \approx_w \overline{c}$ *holds.*

Case 4. Suppose e is an input event of some web service w_i *and* $\langle e\rangle|_w = \langle e\rangle$. *Since* $c \approx_w c'$ *we know* $Q_i \approx_w Q'_i$. *Since* $e \in Q_i$ *there is an* $e' \in Q'_i$ *with* $e \approx_w e'$. *Hence,* $s_i \approx_w s'_i$ *implies that there is a* δ *of the form* $\delta' \circ \langle e'\rangle \circ \delta''$ *such that* $s'_i \rightarrow_\delta \overline{s}'_i$. *Similar to the cases above, the web service* w_i *can execute* δ *locally except that some output invisible to w may be written to some input buffers. Therefore,* $c' \rightarrow_\delta \overline{c}'$ *results in* $\overline{s}'_j = s'_j$ *for all* $j \neq i$ *and* $\overline{s_i}' \approx_w s'_i$ *while* $\overline{Q}_j \approx_w \overline{Q}'_j$ *holds also for all* $1 \leq j \leq m$. *Since* δ *does not contain any output event visible for w* $\overline{M}' \approx_w M' \approx_w M = \overline{M}$ *holds. Thus* $\overline{c}' \approx_w \overline{c}$ *holds.*

6. Web Service Implementation

The definitions 10 and 13 formulate sufficent criterions to ensure the privacy of the data used in a web space. However, the problem remains of how to implement web services for which we can guarantee that they satisfy these conditions. Furthermore, considering dynamic service composition [4, 28, 14] these conditions have to be checked online while processing of a web service request. To simplify matters, we restrict now the way web services can communicate with each other and suppose a specific communication protocol for the implementation. We assume that a web service will request another service by sending one particular message to the other web service and will also receive the result within a single message from the other web service.

OWL-S [19] is a common language to specify web services and provides typical features of a programming language (allowing for sequences, if-then-else, loops, etc.) [1, 18] to implement web services. In order to be used by other web services, a web service exports both its service description and its security policies regulating the distribution of provided data. These policies specify the requirements on the sticky policies for all the input data that are necessary to execute the service and also predicts the sticky policies that will be attached to the resulting data that will be sent back to the calling web service. Since the sticky policies of output data typically depend on the policies for the input, we use polymorphic policies to represent the dependencies between input and output policies.

Enforcing the access control restrictions of Definition 10 is rather easy. Before a message is sent to another web service its sticky policies could be checked. Only if the target web service satisfies the policies of the message then a web service would deliver the message to its designated receiver. However, to enforce the information flow control requirements of Definition 13 we have to analyse the information flow inside the program anyway. Rather than checking the access control restrictions during runtime the analysis of the information flow restrictions will reveal any violations of this property statically.

We base our approach [11, 12] on information flow control in web services on techniques in the areas of language-based security and static program analysis (cf. for instance [22, 23]). The idea is to annotate each expression in the program with a security type. Security is enforced by type checking. A type calculus regulates how types have to be propagated along the constructs of the programming language. Then, a system is secure if it can be typed according to the calculus in a consistent way.

Types are attached to all constructs of the programming language that store or provide information (like, for instance, locations, parameters or variables) and typing rules are formulated that specify the consequences to the involved types once any particular statement of the programming language is executed. The programming language used, for instance, in [27] is a simplified sequential programming language which provides conditionals, loops, and procedures. Since we do not use constructs of OWL-S concerned with concurrency[1], we are able to translate OWL-S program specifications into this particular programming language.

Variables are used to store information. They need clearances to hold classified information. Whenever we want to store a value `t` into a variable x, e.g. by an assignment like `x := t` then the clearance of x has to be higher or equal than the classification of `t`. On the other hand the clearance of a variable to store information determines also its classification once we read the content of this variable. For instance, if we assign x to another variable y by `y := x` then the classification of y is as high as its clearance.

Besides the problem of assigning classified data to uncleared variables there is also the threat of an indirect information flow. Suppose that x is a variable with a high a high classification and y has a low clearance, then a program like `if x = 1 then y = 1 else y = 0;` would allow a user who only sees the value of y to deduce the value of x. To avoid this situation, security types are also assigned to each program fragment (commands or sequences of commands). This type denotes the minimum of the clearances of all variables which are changed inside the fragment. Obviously, we are unable to determine whether the fragment has been executed if we only inspect variables with clearances less than this minimum (because they are all unchanged by the fragment). Hence, we demand that the clearance of x occurring in the guard of the conditional `if...then...else` has to be a lower bound of the security types of both branches in the conditional to ensure the security of the conditional.

Following these ideas each programming construct constitute a set of constraints on

[1] I.e. our underlying service planer does not use this constructs in the first place

the security types of the occurring variables and data. In [11] we extended an approach of Volpano and Smith to cover also web service calls. In this approach web services are treated similar to a procedure except that they are treated as first class citizens having (similar to variables) their own clearances. The call of a web service has to be guarded by a check whether the input to be provided to the service lies within its clearance. Therefore we introduced a new phrase type

$$\tau\ proc(\tau_{in,1},\ldots\tau_{in,n},\tau_{out,1},\ldots\tau_{out,n'})\ \tau'$$

to type web services. This call corresponds to two messages one containing the request for the service and the other delivering the result. First, the phrase denotes that the call of the web service has type $\tau\ cmd$ which means that both messages obtain τ as the sticky policy sp_0. Second, the input has the classification $\tau_{in,1},\ldots,\tau_{in,n}$ (corresponding to $sp_1,\ldots sp_n$ of the first message) and the resulting output requires the clearing $\tau_{out,1},\ldots,\tau_{out,n'}$ (corresponding to $sp_1,\ldots,sp'_n$ of the second message). Additionally, τ' defines the clearance of the web service from the caller's point of view. It is crucial to understand the difference between the classification $\tau_{in,1},\ldots\tau_{in,n}$ of the input and the clearance τ' which the calling web service assigns to the called web service. In general web services are polymorphic in their types; for example, $\tau_{in,1}$ and $\tau_{out,1}$ may be type variables that are related by the constraint $\tau_{in,1}=\tau_{out,1}$. In this case the output would require the same clearance as the classification of the input. An example would be a web service that simply copies its input to the output. Thus $\tau_{in,1}$ and $\tau_{out,1}$ reflect the constraints on the security types considering the computation inside the called web service while τ' reflects the trust the calling web service has in the called web service.

7. Conclusion

We presented an approach to ensure privacy of data within web service communication and used language-based information flow control to ensure the privacy of this data for dynamically composed web services. We provided criteria for individual web services to ensure that the visible behaviour of web services will not disclose any private or confidential information.

However, in general the problem of detecting information flow between different parts of a program is undecidable. Therefore, approaches in language-base security based on an efficient type checking are incomplete, i.e. they may claim an information flow which doesn't exist in reality. However, a variety of different security type systems have been developed in the last decade[2] that advance in reducing such false positives by enhancing the modeling of how information flows in constructs of a programming language. They also improve our abilities to enforce the privacy of our data.

The approach of verifying dynamically composed web services presumes that the planning of the web service is done without knowing or using private data and is only based on generic service desriptions. Otherwise the planning could incorporate confidential data into the design of the program. Suppose, for instance, there is a web service implementing a timer server that has no clearances to see any information about the health status of Peter

[2] for instance [17, 22] to name only two of them; cf. [23] for a detailed description

Miller. However, the pharmacy could encode such information into the number of calls of this time server. Thus the time server has only to count how often it is called by the pharmacy to compute, for instance, an index in a list of diseases (commonly known by him and the pharmacy). Since the calls of the time server do not contain any data the type calculus would not detect such breaches of security. In order to allow for the use of private data also during the planning phase one has to book keep the information used to create the different parts of the plan and upgrade the program fragments representing these parts according to the necessary clearances of the used information. Hence using a confidential information to generate a program fragment, this program fragment needs at least the clearance H which in particular means that it cannot send any messages to uncleared services in between.

References

[1] Ankolekar, A.; Huch, F.; Sycara, K. P.: Concurrent semantics for the web services specification language DAML-s. In: *Coordination Models and Languages,* 2002.

[2] Bell, D. E.; LaPadula, L.: *Secure computer systems: Unified exposition and multics interpretation.* Technical Report MTR-2997, MITRE, 1976.

[3] Brison, J.; Martin, D.; McIlraith, S. I.; Stein, L. A.: *Agent-based composite services in DAML-S: The behavior-oriented design of an intelligent semantic web.*2002.

[4] Bryson, J; Martin, D; McIlraith, S. I.; Stein, L. A.: Agent-based composite services in DAML-S: The behavior-oriented design of an intelligent semantic web. In: *Web Intelligence*, Zhong, N.; Liu, J.; Lao, Y.; editors, Springer Verlag, 2002.

[5] Chen, H.; Joshi, A.; Finin, T.: Dynamic service discovery for mobile computing: Intelligent agents meet Jini in the aether. *Cluster Computing*, 2001, 4(4), 343–354.

[6] Dulay, N.; Damianou, N.; Lupu, E.; Sloman, M.: A policy language for the management of distributed agents. In: *Agent Oriented Software Engineering (AOSE-2001)*, Springer, LNCS 2222, 2001, 84–100.

[7] Fu, X.; Bultan, T.; Su, J.: Conversation Protocols: A Formalism for Specification and Verification of Reactive Services. In: *Proceedings of Implementation and Application of Automata*, **2759**, 188-200. Postprint available free at: http://repositories.cdlib.org/postprints/809

[8] Goguen, J. A.; Meseguer, J.: Security Policies and Security Models. In: *Proceedings of the IEEE Symposium on Security and Privacy,* Oakland, CA, USA, 1982, 11–20.

[9] Goguen, J. A.; Meseguer, J.: Inference Control and Unwinding. In: *Proceedings of the IEEE Symposium on Security and Privacy*, Oakland, CA, USA, 1984, 75–86.

[10] Hutter, D.; Mantel, H.; Schäfer, I.; Schairer, A.: Security of multi-agent systems: A case study on comparison shopping. In: *Journal of Applied Logic. Special Issue on Logic Based Agent Verification*, Article in press, doi:10.1016/j.jal.2005.12.015, 2006.

[11] Hutter, D.; Volkamer, M.: Information flow control to secure dynamic web service composition. In: *Proceedings of the 3rd International Conference on Security in Pervasive Computing, SPC-2006*, York, UK: Springer-Verlag, LNCS 3934, 2006.

[12] Hutter, D.; Volkamer, M.; Klusch, M.; Gerber, A.: Provably secure execution of composed semantic web services. In: *International AAMAS-Workshop on Privacy and Security in Agent-based Collaborative Environments, PSACE-2006*, Hakodate, Japan, 2006.

[13] Karjoth, G.; Schunter, M.; Waidner, M.: Platform for enterprise privacy practices: Privacy-enabled management of customer data. In: *Privacy-Enhancing Technologies*, R. Dingledine and P. Syverson, editors, Springer, LNCS 2482, 2003, 69–84.

[14] Klusch, M.; Gerber, A.; Schmidt, M.: Semantic web service composition planning with owls-xplan. *1st Intl. AAAI Fall Symposium on Agents and the Semantic Web*, 2005.

[15] Mantel, H.: Possibilistic Definitions of Security – An Assembly Kit. In: *Proceedings of the IEEE Computer Security Foundations Workshop*, Cambridge, UK, 2000, 185–199.

[16] McLean, J.: Proving noninterference and functional correctness using traces. *Journal of Computer Security*, 1992, 1(1), 37–57.

[17] Myers, A.; Liskov, B.: A decentralized model for information flow control. In: *ACM Symposium on Operating System Principles*, 1997.

[18] Norton, B.; Foster, S.; Hughes, A.: A compositional operational semantics for owl-s In: *2nd International Workshop on Web Services and Formal. Methods (WS-FM 2005)*, 2005.

[19] *OWL ontology web language, w3c standard technical recommendation*, http://www.w3.org/TR/2003/WD-owl-ref-20030331/.

[20] Patwardhan, A.; Korolev, V.; Kagal, L.; Joshi, A.: Enforcing policies in pervasive environments. In: *Mobile and Ubiquitous Systems, MobiQuitous-2004*. IEEE Computer Society, 2004, 299–308.

[21] Roscoe, A. W.; Goldsmith, M. H.: What is intransitive noninterference? In: *Proceedings of the 12th IEEE Computer Security Foundations Workshop*, Mordano, Italy, 1999, 228–238.

[22] Sabelfeld, A.: *Semantic models for the security of sequential and concurrent programs*. Ph.D. dissertation, Chalmers University of Technology and Gothenburg University, Sweden, 2001.

[23] Sabelfeld, A.; Myers, A.: Language-based information-flow security. In: *IEEE Journal on Selected Areas in Communications*, 2003, 21(1).

[24] Servinc, P. E.; Basin, D.; Olderog, E.: Controlling access to documents: A formal access control model. In: *International Conference on Emerging Trends in Information and Communication Security, ETRICS-2006.* Springer-Verlag, LNCS 3995, 2006.

[25] Sheshagiri, M.; desJardins, M.; Finin, T.: A planner for composing services described in DAML-S. In: *Proceedings of AAMAS 2003 Workshop on Web Services and Agent-Based Engineering,* 2003.

[26] Volpano, D. M.; Smith, G.: A sound type system for secure flow analysis. In: *Journal of Computer Security*, 1996, 4(3), 167–187.

[27] Volpano, D. M.; Smith, G.: *A type-based approach to program security.* In: TAPSOFT, 1997, 607–621. [Online]. Available:
citeseer.ist.psu.edu/volpano97typebased.html

[28] Wu, D.; Parsia, B.; Sirin, E.; Hendler, J.; Nau, D.: Automating DAML-S web services composition using SHOP2. In: P*roceedings of the 2nd International Semantic Web Conference (ISWC2003)*, Sanibel Island, Florida, USA, October 2003, 20–23.

[29] Zakinthinos, A.; Lee, E. S.: A General Theory of Security Properties. In: *Proceedings of the IEEE Symposium on Security and Privacy,* Oakland, CA, USA, 1997, 94–102.

In: Long-Term and Dynamical Aspects of Information... ISBN 1-60021-912-8
Editors: Andreas U. Schmidt et al, pp. 45-59 © 2007 Nova Science Publishers, Inc.

Chapter 4

GEOGRAPHICALLY SECURE ROUTING FOR MOBILE AD HOC NETWORKS: A CROSS-LAYER BASED APPROACH

André König,[*] Matthias Hollick[†] and Ralf Steinmetz[‡]
Multimedia Communications Lab (KOM)
TU Darmstadt, Merckstrasse 25, 64283 Darmstadt

Abstract

The growth of mobile and wireless communications has also raised various concerns with respect to information-security. Recently, the design of secure protocols for wireless LANs has been placed under scrutiny to ensure the authenticity, integrity, and confidentiality of the communication. However, these protocols mostly focus on the protection of the ongoing communication only and neglect long-term security goals. Attack vectors against such long-term security goals include, but are not limited to adversaries that are eavesdropping the wireless communication to perform a post hoc cryptanalysis (possibly years after the data-capture). This paper introduces the paradigm of geographically secure routing to address the aforementioned shortcomings of existing security schemes. In particular, we propose a solution to transmit sensitive information in a mobile ad hoc network such that the routes are restricted to trustworthy nodes. Our scheme is able to prevent eavesdropping of or tampering with data in-flight, thus, thwarting passive and active attacks and supporting the long-term security of the system. We implement our solution using a cross-layer approach that builds on existing ad hoc routing protocols and maintains the compatibility with these protocols. A simulation study shows the feasibility of our approach and validates the proposed solution.

[*]E-mail address: Andre.Koenig@KOM.tu-darmstadt.de
[†]E-mail address: Matthias.Hollick@KOM.tu-darmstadt.de
[‡]E-mail address: Ralf.Steinmetz@KOM.tu-darmstadt.de

1. Introduction

1.1. Motivation

The growth of mobile and wireless communications has also raised various concerns with respect to information-security. Several breaches in security protocols for wireless communication systems based on IEEE 802.11 Wireless LAN, Bluetooth, or GSM have recently been reported. As a result, the design of secure protocols in the wireless communication domain has been placed under scrutiny to ensure the authenticity, integrity, and confidentiality of the communication. Mechanisms and protocols such as WEP, WPA, IEEE802.11i on lower layers are combined with higher layer mechanisms such as IPSec and TLS to meet the security demands. Since these approaches are all based on currently available encryption mechanisms, they are potentially vulnerable regarding their long-term security. Attack vectors against such long-term security goals include, but are not limited to adversaries that are eavesdropping the wireless communication to perform a post hoc cryptanalysis. The mathematical problems that build the fundament of today´s encryption algorithms might be solved in affordable time in the future if we consider improvements in computing power, e.g., based on advances in quantum computing [1]. The breaking of the DES algorithm, as described in [8], is one example of the past that surely will show up again. In summary, the application of traditional security schemes in wireless networks opens up various problems that demand for novel solutions. This paper introduces the paradigm of geographically secure routing to address the aforementioned shortcomings of existing security schemes. In particular, we propose a solution to transmit sensitive information in a mobile ad hoc network such that the routes are restricted to trustworthy nodes. Our scheme is able to prevent eavesdropping of or tampering with data in-flight, thus, thwarting passive and active attacks and enhancing the long-term security of the system.

1.2. Contribution

In this work we investigate novel mechanisms to support the long-term security coupled to mobile/wireless ad hoc networks. We develop a solution to provide geographically secure routes on top of legacy routing protocols such as Dynamic Source Routing (DSR). Our solution follows a cross-layer approach to control the routing protocol from application level. Our main contribution is the design, implementation, and experimental validation of a geographically secure routing scheme. We follow a cross-layer approach to couple the application layer (which specifies the security requirements) with the network layer (here: legacy ad hoc routing protocols). Our solution aims to enhance the long-term security of corporate ad hoc network deployments.

1.3. Outline

The remainder of this paper is structured as follows. In Section 2., we describe the scenario underlying our work. We perform a security analysis and derive the requirements for possible solutions. Our novel approach towards geographically secure ad hoc routing is precisely described in Section 3.. In particular, we motivate the choice of a cross-layer architecture for our solution and give a description of the routing control interface that includes all necessary protocol extensions. As a proof-of-concept we perform a simulation study to validate the operation of our routing mechanism. We outline the design of the simulations and analyze and present selected results in Section 4.. Finally, we summarize our work in Section 5..

2. Scenario

Within our scenario we consider a research site with employees and visitors carrying mobile communication devices. Visitors have to be considered as non-trustworthy and are therefore assumed to be restricted to a designated area within the research site. Employees are considered as trustworthy. For them, no restrictions with respect to movement are made.

From the technical point of view, trusted employee devices are further divided into devices whose functionality can be changed with small effort (like notebooks or PDAs running open source operating systems) and devices where this is not (easily) possible (like network printers with proprietary operating systems).

Our approach is based on an extension of the functionality of nodes. Nodes, whose functionality we extend, are further called *extended nodes*. Nodes with unchanged functionality are referred to as *standard nodes*.

For communication between any two involved devices, we use a mobile ad hoc network. The exchanged information is classified into confidential and non-confidential data flows. To setup the required routes, the deployed routing protocol is DSR [5].

From the described scenario, we can extract four possible communication cases, which are shown in Figure 1:

- Employee nodes exchanging confidential information
- Employee nodes exchanging non-confidential information
- Employee node and visitor node exchanging non-confidential information
- Visitor nodes exchanging non-confidential information

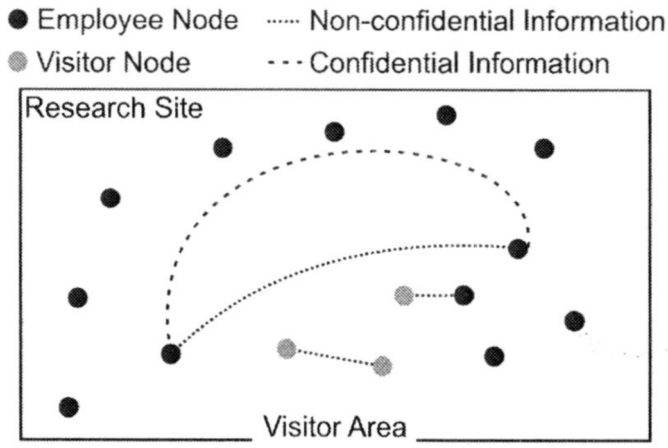

Figure 1. Scenario with schematic end-to-end communication relationships.

2.1. Security Analysis

Various attacks that are inherently possible in mobile ad hoc networks due to their infrastructureless nature have been identified so far [6]. These can be classified into active and passive attacks. In general, the intention of an active attack is to change the data flow in a mobile ad hoc network. For this, active attacks require changes in the behavior of the deployed routing protocols to achieve the desired effects. Thus, nodes that perform active attacks, can be detected (and located) by an intrusion detection system, as proposed for example in [7]. In contrast to this, passive attacks like traffic analysis or eavesdropping of specific communications do not require the attacker to change the routing protocol and have no direct effect on the behavior of the mobile ad hoc network as a whole. In fact, passive attacks do not require the node to transmit any information, what makes it (nearly) impossible to detect passive attackers. Active attack mechanisms may (but do not have to) be used in combination with eavesdropping in order to make the result even worse (or better from the perspective of the attacker).

Today, data encryption is the method of choice to prevent that information collected during an eavesdropping attack can be exploited by the attacker. In view of long-term security, state of the art encryption mechanisms could fail in a few years, giving the potentially malicious visitor in our scenario the chance to reveal the secrets once collected. If we further consider public key infrastructures [2], where a private/public key pair is usually used for a long period of time, it might be possible for an adversary to compute the private key from collected data and to later still misuse this knowledge. From this perspective, a feasible way to keep information confidential in the future with today's techniques is to keep it away from unauthorized persons.

For our given scenario, we expect all attacks, whether they are of active or passive nature, to be restricted to the visitor area. The consequence is that confidential information should not enter the visitor area. This is shown in Figure 1 where end-to-end communication relationships are sketched.

2.2. Resulting Requirements

In the cases of non-confidential communication, the visitor nodes should be used for hop by hop information forwarding in order to provide the expected connectivity. A respective route in our scenario is shown in Figure 2(a). This route will most likely be chosen by DSR since it is the shortest with respect to number of hops.

In the case of a confidential communication between employee nodes, a visitor node should (for the reasons described in the previous section) not be part of a path between employee nodes. So at the first stage, we need the ability of an explicit user interaction or an implicit policy mechanism within the utilized application to inform the routing process whether or not the information to transmit is confidential.

For confidential communications, we have to establish routes which bypass the visitor area. Figure 2(b) shows a route in our scenario that meets this restriction.

When we take a look at node X in Figure 2(b), we see that due to its proximity to the visitor area, its transmission would also reach unauthorized nodes. Even though X is a trusted employee node, it should at its current position not be used for forwarding confidential information.

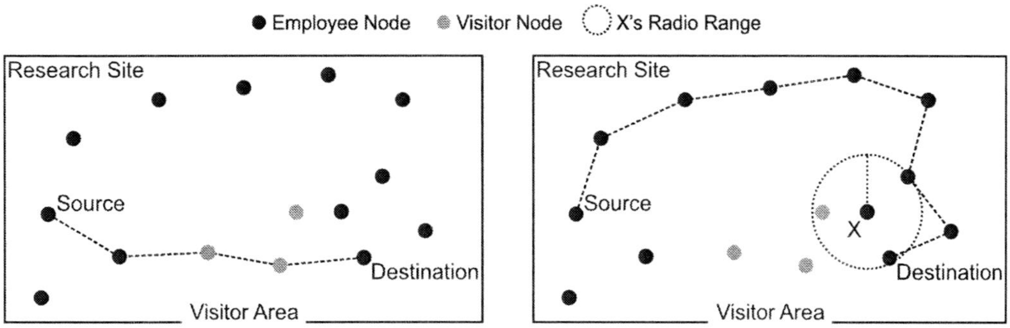

(a) Non-confidential route (b) Confidential route

Figure 2. Confidential versus non-confidential route.

For the decision whether an employee node may be contained in a route that is used for confidential communication, information about its position and its radio range has to be available.

With respect to the subclassifications of employee devices, we have to assume that connectivity decreases if we restrict routes to employee nodes whose routing functionality we extended to distinguish between confidential and non-confidential communication and to handle position information (extended nodes). To overcome this, we allow a route to contain a certain number of non-extended employee nodes (standard nodes) between any two adjacent extended nodes. The endpoints of a route, that is sender and receiver, are expected to be extended nodes.

The properties of our scenario and the requirements can be summarized as follows:

- Devices are classified into employee devices and visitor devices
- Visitor devices are restricted to the visitor area
- Employee devices are further classified into extended nodes and standard nodes
- End points of communications are extended nodes
- An information exchange between application and routing process is needed
- Knowledge of the position and the radio range of extended nodes has to be available

3. Approach

To achieve the desired functionality as described in the previous section, we have to take influence on the vertical control flow within one node (from application to routing process) as well as on the horizontal data flow between two nodes. In the following, we describe the required extensions.

3.1. Cross-Layer Architecture

For reasons of compatibility, we base our approach on the well established Internet model with its strictly separated layer architecture, as shown in Figure 3(a). To exchange the necessary information between the application and the routing process, we add a cross-layer extension similar to the design proposed in [3] in a two step process.

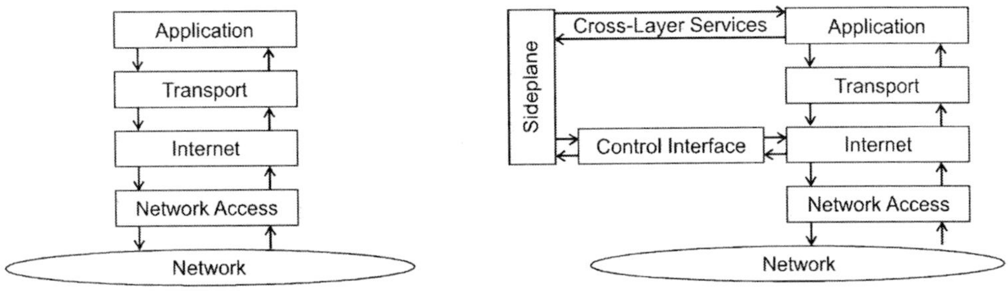

(a) Internet layer model

(b) Cross-layer design

Figure 3. Extension of the Internet layer model.

In the first step an additional control interface which will be described in detail in the next section is attached to the network layer. This way, the desired influence on the routing process becomes possible. A similar approach with the aim to rewrite routing tables in order to optimize Gnutella networks is presented in [4].

Step two adds an orthogonal sideplane, which offers the service primitives for cross-layer communication. A draft of the resulting architecture is given in Figure 3(b).

The sideplane is organized as a lightweight data structure containing *(name, value)* tuples. Services are offered to add and change tuples, as well as to register a process to be informed about changes in a specific tuple. With respect to our scenario, an application adds and changes tuples as for example the required position information of the node *("NodePosition", GPS coordinates)* and the visitor area *("VisitorArea", Polygon)*. The control interface of the network layer registers for changes in both tuples. The information is then used to influence the routing process respectively.

3.2. Routing Control Interface

To stay compatible with nodes that run a standard DSR protocol, we leave the DSR header unchanged. The necessary information is contained in an additional header that follows the DSR header. We assign the header number *253* that is reserved by IANA for experimentation and testing. Thus the value of the *next header* field of the DSR header (which itself has not been assigned a fixed number yet) is *253* and points to our additional header. The resulting MAC frame is shown in Figure 4.

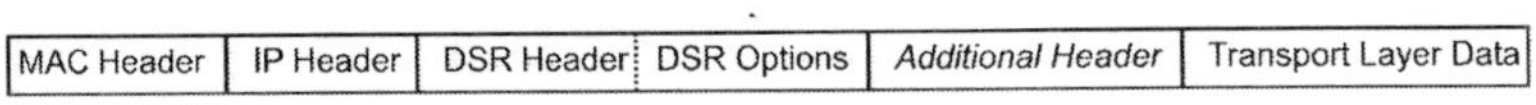

Figure 4. MAC frame with additional header.

Nodes that run a standard DSR protocol simply ignore the additional header, whereas nodes that are equipped with our extension can read and evaluate the contained information. We define three header formats for the three phases of a communication which are route request, route reply and data transfer:

3.2.1. Route Request

To provide security at the earliest possible point in time, we already demand the route request not to reach the visitor area. This way, we prevent visitors to perform traffic analysis or (if able to handle our extension) to pretend a position outside the visitor area with the aim to be included in a confidential route.

For the route discovery phase, the additional header contains the following information:

Next Header This is used to determine the transport layer protocol as for example TCP or UDP.

TTL The time to live for this route request. This field is decreased at each extended node by the number of hops that where traversed since the previous extended node. If we allow a route to only consist of extended nodes, TTL will be decreased by one at each (extended) node. If standard nodes are allowed to be situated between extended nodes, TTL is decreased respectively.

Max. intermediate standard nodes This field specifies the maximum amount of standard nodes that may be situated between two adjacent extended nodes. The TTL of the IP header is set to this value at every extended node. This way, the broadcast of a route request that will be done by standard DSR nodes is restricted to the desired amount of standard nodes between adjacent extended nodes.

Expected Replies The amount of expected route replies which will be of relevance for our multipath approach as a part of our future research.

Sequence Number This field is reserved for our multipath approach.

Header Length The overall length of the additional header. The length is not fixed, since the following field contains a flexible description of the position and the shape of the visitor area.

Restricted Area A polygonal model of the visitor area.

Figure 5 depicts the resulting structure of the additional header for the route request phase.

Next Header	TTL	Max. Intermediate	Expected Replies
Sequence Number		Header Length	
Restricted Area			

Figure 5. Header format for route request.

3.2.2. Route Reply

During the route reply phase, each extended node appends its current geographical position and its position in the recorded list of hops in the DSR options to the extended header. The resulting header format is shown in Figure 6.

<table>
<tr><td>Next Header</td><td>Reserved</td><td>Header Length</td></tr>
<tr><td colspan="2">Position in DSR Hop List</td><td>Reserved</td></tr>
<tr><td colspan="3">GPS coordinates</td></tr>
</table>

Figure 6. Header format for route reply.

Based on this information, the source evaluates the degree of security, a route can offer. As an example, we assume that one standard node is situated between two extended nodes. For a worst case scenario, we further assume each extended node to be as close to the visitor area as it is allowed by its radio range.

If the distance between the extended nodes then converges to the sum of their radio ranges, the transmission of the intermediate standard node can not reach the visitor area, as shown in Figure 7(a).

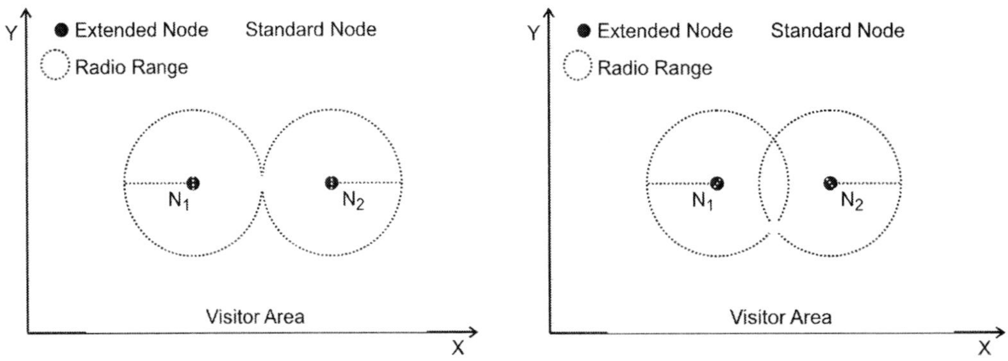

(a) Secure route (b) (Possibly) unsecure route

Figure 7. Route with one intermediate standard node.

If, on the other hand, the distance between the extended nodes is smaller than the sum of their radio ranges, the transmission of the intermediate standard node may well reach the visitor area. Regarding this, a quantitative assertion about the security of a route becomes possible. The possibility that the transmission of an intermediate node reaches the visitor area can be calculated from the knowledge of the position of the two neighboring extended nodes. Figure 7(b) depicts the worst case of a (possibly) unsecure situation. We have to notice, that the intermediate standard node could be situated anywhere within the intersecting plane of the radio ranges of the extended nodes.

3.2.3. Data Transfer

During the phases of route request and route reply, we established a route, that meets our security requirements. Since we are confronted with mobile devices, the route has to be maintained with respect to security during the data transfer phase.

To obtain the required flexibility for our further work on more unrestricted scenarios, we define a header for the data transfer phase, that may be used optionally when the scenario (and thus the restrictions with respect to security) changes. The header therefore contains for each extended node in the DSR hop list a description of an area, where the node is allowed to detain. If a node leaves this area, it has to stop forwarding messages for the respective communication.

Figure 8 shows the additional header that is used during the data transfer phase.

Next Header	Reserved	Header Length
Position in DSR Hop List		Reserved
Permitted Area		

Figure 8. Header format for data transfer.

4. Simulation and Evaluation

We apply the methodology proposed by Jain [9] for the experimental analysis of our approach and adopt the individual steps to our scenario. The methodological steps can be summarized as follows:

- Definition of the system, goals, and services
- Selection of the metrics
- Definition of the parameters to study
- Selection of the factors/elements of the parameter set
- Choice of the evaluation technique
- Selection of the workload
- Design of the individual experiments
- Analysis and interpretation of the obtained data
- Presentation of the results

4.1. Simulation Setup

A detailed description of the simulation setup regarding the dimensions is shown in Figure 9. We consider a research site of 3000 meters width and 2000 meters height. The visitor area is situated at the center of the bottom line with a width of 1000 meters and a height of 500 meters. The radio range of 250 meters is equal for each node within our scenario. No packets are lost during transmission.

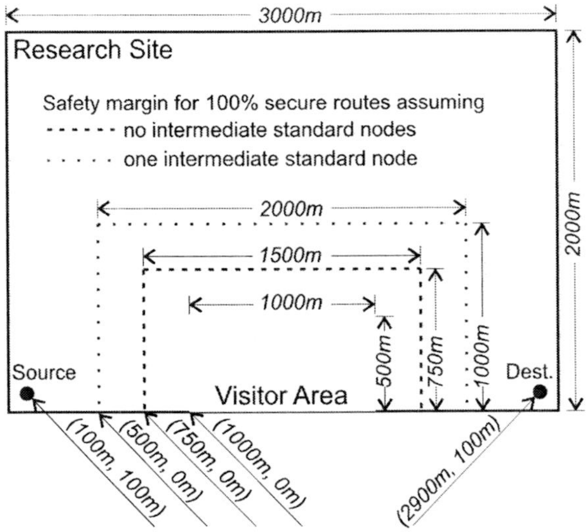

Figure 9. Dimensions of our scenario.

To obtain a worst-case scenario for our approach, we place the source and the destination in the lower left and the lower right corner of the research site. If successful, our approach discovers a route which bypasses the visitor area, whereas we expect standard DSR to discover a route straight through the visitor area, as we already drafted in Figure 2.

The utilized simulation tool is an adapted version of JiST/SWANS [10], a Java based open source discrete event simulator for wireless ad hoc networks. JiST/SWANS has shown to be easily extendable and to provide good scalability and performance.

In a first step, we compare the connectivity of standard DSR to the connectivity of our approach. As a metric for this we use the fraction of successful route requests out of the number of total route requests. A statistical mean value is determined during 1000 simulation runs with random node placement and one route request each. Sender and receiver are placed at the fixed positions as shown above. Each extended node with a position outside the inner safety margin shown in Figure 9 (so at least the sending range of 250 meters away from the visitor area) is allowed to be contained in a route.

For this evaluation, the parameters for each 1000 runs are

- the total number of nodes in the scenario,
- the fraction of extended nodes, and
- the number of intermediate standard nodes between two adjacent extended nodes.

In the second step, we quantitatively evaluate the degree of security that is reached, if we allow one intermediate standard node to be situated between two adjacent extended nodes. For this, we choose a fixed parameter set, that showed to achieve 100% connectivity in the first step of evaluation described above. We then introduce a second safety margin, as shown in Figure 9. No extended node that is situated within this area is allowed to forward messages. It is obvious that if this second safety margin then has the size of 250 meters (measured from the inner margin), no messages reach the visitor area, since the sending range of an extended node and a following standard node can at most be 500 meters.

For the second step, the parameter that is considered in simulation is the width of the outer safety margin. Again, we perform 1000 simulation runs with random node placement and one route request each. Sender and receiver stay fixed at the positions shown in Figure 9. To check whether a discovered route is in reception range of the visitor area, we modeled listening nodes for JiST/SWANS. These are placed with a distance of 100 meters along the boundaries of the visitor area. The listening nodes do not forward any messages and thus have no effect on the route discovery process. The metric to measure the security of a route is the fraction of route request messages that are received by the listening nodes around the visitor area out of the total amount of route request messages that are sent during the route request phase.

4.2. Simulation Results and Evaluation

To graphically show that our approach works in general, Figures 10 and 11 provide screenshots of the route request phases of our approach and DSR. The visualization is done with a graphical monitoring tool for the JiST/SWANS simulator which has been developed at our institute. We model a static simulation setup, that is similar to the scenario as depicted in Figure 2.

In Figures 12 and 13, the results of the first step of the evaluation of our approach are shown. Figure 12 depicts the evaluation of the connectivity of our approach and of DSR. For this evaluation we used homogeneous setups, which consist either of standard DSR nodes or of extended nodes.

As we expected, the price for a secure route is a decrease in connectivity of our approach compared to standard DSR. The mean decrease in connectivity for our scenario shows to be approximately 20%. The elaboration of the reason for this will be part of our future work.

For the evaluation shown in Figure 13, a heterogeneous setup of standard and extended nodes is considered. Depicted is the connectivity of routes which only consist of extended nodes and of routes which may contain one intermediate standard node between adjacent extended nodes. We simulate a random distribution of 400 nodes and stepwise increase the fraction of extended nodes.

In Figure 14 the results of the second step of our evaluation are presented. We simulate a setup with 200 standard nodes and 200 extended nodes. One intermediate standard node is allowed to be situated between two adjacent extended nodes. The size of the outer safety margin is increased stepwise.

Like presumed, the degree of security of a route decreases if we allow standard nodes to be contained in a route and along with this reduce the size of the outer safety margin. The explanation for this observation is that like DSR, our approach will most likely find

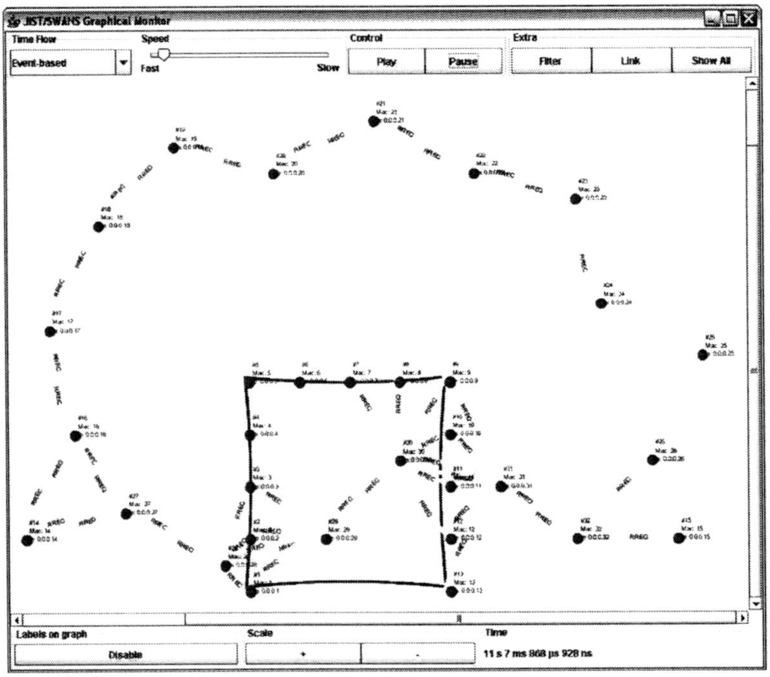

Figure 10. Standard DSR route request.

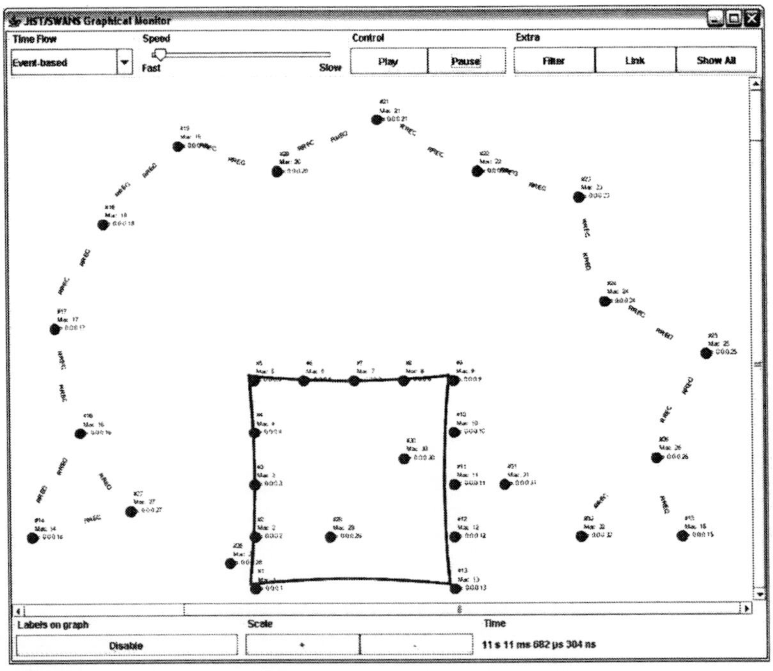

Figure 11. Restricted route request.

the shortest route with respect to number of hops. This route is most likely the one with the shortest distance to the visitor area. Therefore, as shown in Figure 7, the transmission

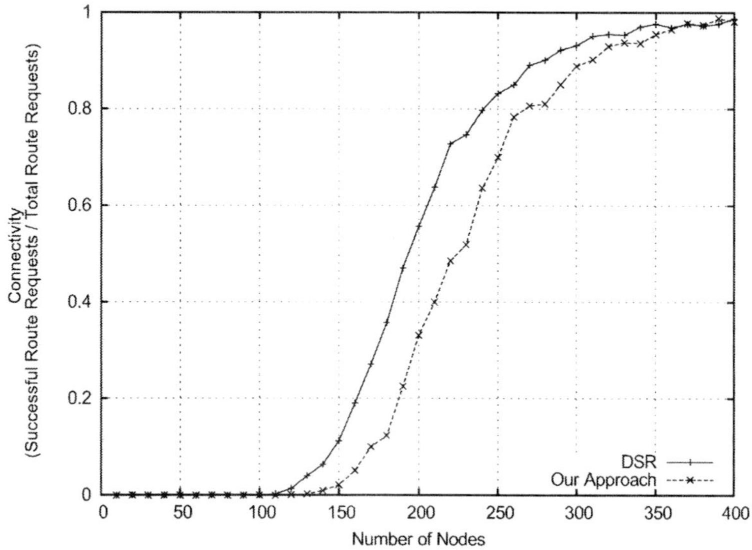

Figure 12. Connectivity of DSR and our approach.

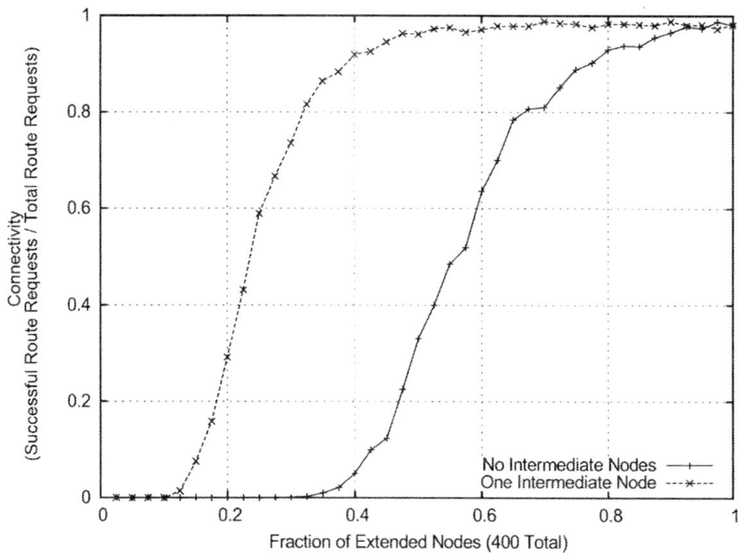

Figure 13. Connectivity for routes with none and one intermediate standard node.

of an intermediate standard node can reach the visitor area, if the distance between the neighboring extended nodes is smaller than the sum of their radio ranges. In this evaluation, our approach always performs better than standard DSR. For our scenario, DSR shows to have a constant rate of 50% intercepted route request messages out of the total amount of transmitted route request messages.

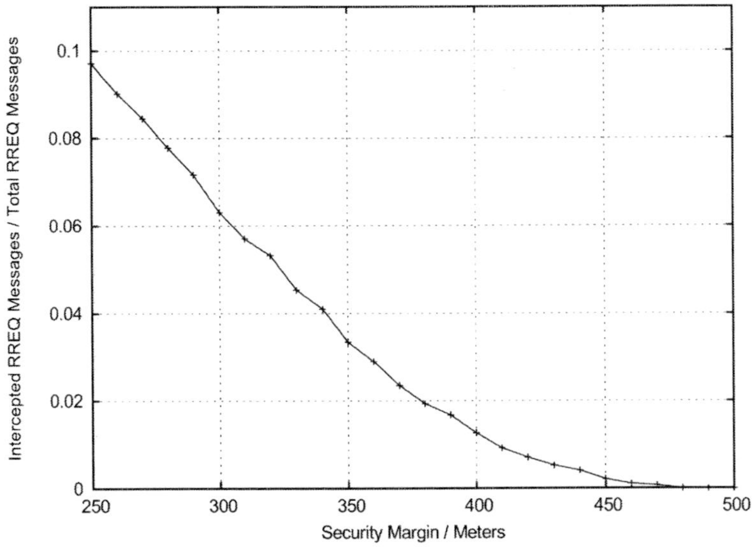

Figure 14. Quantitative security of our approach.

5. Summary and Outlook

Keeping away confidential information from unauthorized persons turns out to be a feasible support for today's available security mechanisms. This especially holds for scenarios with a demand for long-term security.

Through a cross-layer extension of the existing Internet layer model, we achieve interaction between the different layers. This enables us to restrict routes in mobile ad hoc networks to nodes which meet our security requirements and therefore to prevent that their transmission reaches non-trustworthy nodes.

Routing with regard to the geographical position of nodes with extended functionality and number of intermediate nodes with unchanged functionality as two degrees of freedom have been our concerns for now. Our scenario quickly reaches a high complexity when we take more than two (visitor, employee) levels of trustworthiness into account. Also a dynamic change of these levels may be necessary, when an employee enters the visitor area. Furthermore, visitors may be allowed to move relatively unrestricted throughout the research site. These more realistic and thus more complex scenarios will be the focus of our future research.

Within these scenarios, also temporal aspects will be part of our research. As an example we plan to delay sending on transport layer in order to prevent a high security level node from transmitting while in proximity to a low security level node. We furthermore will consider the adaptation of the transmission power of nodes to increase the number of possible routing devices and with this further improve connectivity.

References

[1] Brassard, G.: Cryptology Column - Quantum Computing: The End of Classical Cryptography? *SIGACT News*; ACM Press, 1994; Vol. 25, pp 15-21.

[2] Henry, D: Who's Got the Key? In *SIGUCCS '99: Proceedings of the 27th annual ACM SIGUCCS conference on User services*; ACM Press, 1999; pp 106-110.

[3] Conti, M.; Crowcroft, J.; Maselli, G.; Turi, G.: A modular cross-layer architecture for ad hoc networks. In *Handbook on Theoretical and Algorithmic Aspects of Sensor, Ad Hoc Wireless, and Peer-to-Peer Networks*; Wu, J.; Ed.; CRC Press, 2005.

[4] Conti, M.; Gregori, E.; Turi, G.: A Cross-Layer Optimization of Gnutella for Mobile Ad hoc Networks. In *MobiHoc '05: Proceedings of the 6th ACM international symposium on Mobile ad hoc networking and computing*; ACM Press, 2005; pp 343-354.

[5] Johnson, D.B.; D.A. Maltz, D.A.; Hu, Y.: The Dynamic Source Routing Protocol for Mobile Ad Hoc Networks (DSR). *Internet-Draft*; IETF MANET Working Group, 2004.

[6] Wu, B.; Chen, J.; Wu, J.; Cardei, M.: A Survey on Attacks and Countermeasures in Mobile Ad Hoc Networks. In *Wireless Network Security*; Xiao, Y; Shen, X.; Du, D.Z.; Ed.; Network Theory and Applications Vol. 17; Springer, 2006.

[7] Tseng C.-Y.; Balasubramanyam, P.; Ko, C.; Limprasittiporn, R.; Rowe, J.; Levitt, K.: Intrusion detection: A specification-based intrusion detection system for AODV. In *SASN '03: Proceedings of the 1st ACM Workshop on Security of Ad Hoc and Sensor Networks*; ACM Press, 2003; pp 125–134.

[8] Electronic Frontier Foundation: *Cracking DES - Secrets of Encryption Research, Wiretap Politics & Chip Design*. O'Reilly, 1998.

[9] Jain, R.: *The Art of Computer Systems Performance Analysis: Techniques for Experimental Design, Measurement, Simulation, and Modeling*. Wiley-Interscience, 1991.

[10] Barr, R.: *JiST - An efficient, unifying approach to simulation using virtual machines*. PhD dissertation; Cornell University, 2004.

In: Long-Term and Dynamical Aspects of Information... ISBN 1-60021-912-8
Editors: Andreas U. Schmidt et al, pp. 61-71 © 2007 Nova Science Publishers, Inc.

Chapter 5

OPENAMBIENT: A PERVASIVE ACCESS CONTROL ARCHITECTURE

***M. Anisetti*[*]*, C.A. Ardagna*[†]*, V. Bellandi*[‡] *and E. Damiani*[§]**
University of Milan, Department of Information Technology
via Bramante 65, Crema, 26013, Italy

Abstract

Technological improvements of sensing technologies make available a great amount of fine-grained ambient information, which can be exploited to secure pervasive computing applications for critacal domains. In this chapter, we introduce open ambient, a web-based architecture that integrates an access control model and language with personalization and location technologies to provide an ambient-aware access control.

1. Introduction

For a long time, lack of reliable security and privacy solutions has been considered to be a major hurdle toward the development of pervasive computing applications for critical domains such as secure workplaces, healthcare facilities and assisted living environments. Today, an advanced security infrastructure for context-aware and personalized authentication and authorization services in heterogeneous networks is gradually taking shape [14]. This infrastructure will enable large-scale mobility using a variety of mobile devices including SIMs [11, 13] as well as other emergent technologies such as video sensors supporting human posture and face interpretation [15, 31]. In such a scenario, fine-grained ambient information coming from monitoring and surveillance devices is used to enrich context representation underlying advanced access control and security policies. Recent research [8] has proposed a service-oriented architecture capable of preserving privacy and protecting

[*]E-mail address: anisetti@dti.unimi.it
[†]E-mail address: ardagna@dti.unimi.it
[‡]E-mail address: bellandi@dti.unimi.it
[§]E-mail address: damiani@dti.unimi.it

resources, including personal user data. Here, we put forward the idea of integrating traditional access control models and languages, based on the attributes of the requestor, with emergent personalization and location techniques to provide an ambient-aware, service-oriented access control.

The remainder of the chapter is organized as follows. After providing an overview of location algorithms (Section 2.), Section 3. describes our technique for high precision mobile terminal geolocation based on lookup table approach and on our time-forwarding tracking technique. Following, in Section 4., we discuss the context representation problem and in Section 5. we describe *OpenAmbient*, a Web Service-based architecture that incorporates an access control language supporting ambient predicates. Thanks to automatic location-based access control and authentication, OpenAmbient context-aware policies can control access to distributed services and resources focusing on mobile users. Finally, Section 6. gives our conclusions and an outline of future research.

2. Location Algorithm: An Overview

The term *geolocation* is currently used for a variety of techniques aimed at computing and tracking the position of a mobile terminal. Satellite-based positioning like GPS (Global Positioning System), for example, is an interesting option for high-end applications, where precision is of paramount importance. However, GPS bsed location is unfeasible in dense urban areas or inside buildings where satellites are not visible from the mobile terminals. A-GPS is an enhanced version of GPS that differs from standard GPS due to the addition of a further element, the *Assistance Server*. In an A-GPS network, the receiver has limited processing capabilities and communicates with an assistance server having more processing power and accessing a reference network. Note that the GPS system does not deal with geo-referencing, i.e. using the geographical coordinates provided by satellites to find and display the terminal location on a map. When necessary, maps used to display the position of a GPS user must be stored on the GPS receiver itself, with a potentially high impact on hardware requirements. Even leaving aside costs and impact on battery consumption, GPS and A-GPS are therefore not likely to be the key technologies for a number of interesting location-based services (LBSs) such as mobile tracking certification.

Positioning and Motion Tracking System (PMTS) represents a general purpose, low-cost solution for mobile networks. PMTS looks attractive for two major reasons: (i) it exploits only information available to the mobile network itself, and (ii) it does not need additional hardware. Using the cellular network for geolocation is straightforward when it is sufficient to determine the cell where a mobile terminal is located. However, some services (including emergency rescue [1], urban traffic management, tracking patients with some particular pathology, and other security services) require more precise terminal location. All these applications require tracking of mobile terminal position within cells, and consequently must rely on high precision location technology. Current location technologies monitor field power values [2, 3], as received from mobile terminals, as well as propagation time [32]. *Time-based methods,* such as AOA, TOA and TDOA [4], rely on differences in propagation time between a terminal and different base stations whose position is known. They work much better when LOS (line-of-sight) between terminals and base stations is guaranteed. An interesting approach based on the same principle as TDOA is E-OTD [5]

(Enhanced-Observed Time Difference), where the difference in propagation time between the base station, the mobile terminal and a reference measurement station with known co-ordinates is computed by correlating two received signals.
Path-loss techniques, instead, try to estimate the location of a mobile terminal using the attenuation of the signal received, by the terminal itself, from one or more base stations. However, it is well-known that an approach based on path-loss that relies only on information available on terminals [35] or on simple geographic information [34] cannot guarantee high precision, even when advanced propagation algorithms are used.

3. A Technique for High Precision Mobile Terminal Geolocation

In this section, we discuss how high-precision geolocation of mobile terminals can be obtained by relying on a standard GSM cellular network. Namely, we propose a novel technique for high precision location and tracking of mobile terminals taking advantages from ambient information, that can be extracted online from a GIS database of the area of interest in conjunction with advanced predictive filtering, to improve precision in location and tracking. This technique, together with face recognition solutions, is at the base of our OpenAmbient architecture (see Section 5.).

3.1. Lookup-Table Geolocation

Our algorithm is based on the so-called database correlation approach [34]. The position of the mobile terminal is determined by comparing the signal power of the 6 best-serving antennas (as measured by the terminal itself) with the entries of a look-up-table, which contains predicted path-loss for all antennas in the area. Using these prediction values we obtain the matrix shown in Fig. 1. Every row of the matrix represents a single point within the coverage area (expressed in Cartesian x, y coordinates if a 2D representation of area is used or as a latitude and longitude pair for GPS representation). The path loss predictions from *n* base stations to each given point are stored as entries in the row corresponding to that point.

Location		Cell			
x	y	1	2	i	r
x_1	y_1	$E_{1,1}$	$E_{2,1}$	$E_{i,1}$	$E_{r,1}$
x_2	y_2	$E_{1,2}$	$E_{2,2}$	$E_{i,2}$	$E_{r,2}$
⋮	⋮	⋮	⋮	⋮	⋮
x_m	y_m	$E_{1,m}$	$E_{2,m}$	$E_{i,m}$	$E_{r,m}$

Figure 1. Look-up table structure.

Of course, look-up table filling and update can be done only once in a while, when major changes on the area of interest occur[1]. Generally speaking, computing our look-up table for a given area means super-imposing a grid which quantizes field levels. This grid cells needs not be of uniform size; rather, the grid's coarseness can be tuned based on the

[1] In principle, all *Electro Magnetic Field* (EMF) prediction models can be used to compute the look-up table, depending on application requirements.

characteristics of the area of interest and on cost constraints. When locating a mobile terminal the best matching entry in the look-up table is searched. During this process the path loss measurement of the terminal is compared to all entries of the matrix. This comparison can be done based on many different criteria; some interesting ones are presented in [34]. In our system, we used a sum of squared errors between measured path loss M_j for each antenna j and the path loss defined by an entry i in the look-up table $E_{i,j}$ (1). Therefore we have:

$$e = \sum_{j=1}^{r} (M_j - E_{i,j})^2 \tag{1}$$

The location of the mobile terminal is defined as the coordinate of the entry that produces the smallest error e. Of course, this simple, single-point location technique is affected by an intrinsic error, due to discrepancy between the real field strength and its predicted value. The principal causes of error in predicting EMF strength are: *i)* intrinsic model error, e.g. due to irregular skyline; *ii)* imprecision in geographical database. A low precision in the GIS database, for example due to information aging of the map, can greatly reduce the precision of the field strength estimates; *iii)* diverse antenna shapes. Any kind of antenna gain variation can result in a discrepancy between predicted and real field[2]; *iv)* weather conditions. Any variation in weather conditions may produce changes in field strength that are not easy to model. Again, this error is spread all over the area of interest.
For these reasons, we adopted a multi-point approach which produces a variable number n of position estimates depending on a sensibility map analysis. Our *sensibility map* is built based on GIS map information, antennas positions, and number of candidate values, and is used to control the error sensitivity of our multiple-candidates geolocation method [7]. Based on our sensitivity analysis, we decided to use $4-6$ estimates and a number between 3 and 5 antennas for an area of $1200m^2$. Experience has shown that this techniques substantially improves the geolocation correctness w.r.t. the single point method.

3.2. Time-Forwarding Tracking (TFT)

For validating a candidate location determined with look-up table described in the previous section we developed a tracking method that uses a time-forwarding algorithm providing a first-level movement estimate [6]. This algorithm uses m time position estimates to create an acyclic graph. Possible positions of the mobile terminal are represented by graph nodes, while arcs represent motion between them. Every arc of the graph has a weight based on map constraints. Weights can be defined as follows:

$$w = 1 - P_p(V, a|map) \tag{2}$$

The weight w expresses the probability P that the location estimate corresponding to node p is the correct one. This probability depends upon velocity V and acceleration a of the mobile terminal, and it is conditioned by map constraints *map*. By searching the best path on the graph (i.e., the path from time t to time $t{+}m$ with minimum weight) we obtain a filtered position (see Fig. 2). All map and motion constraints are enforced by construction

[2]Note that this type of error is not localized but spread all over the cell.

of the weights. Using this approach the m parameter becomes critical. In fact using long time prevision (i.e., taking a high m) we obtain a "strong trend" prevision that filters out every out-of-trend movements. This result is not acceptable in pedestrian-only areas like a city square where motion can well be chaotic without any prevalent movement trend. On the other hand, using a too small value for m could cause an error dependency which in turn could produce bad results in high trend-correlation area like motorways or one-way streets.

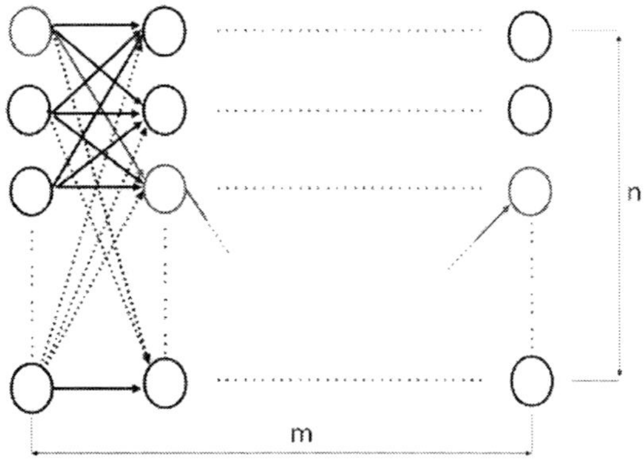

Figure 2. Graph trough example for time-forwarding tracking. In red the selected position node and the selected path, in black the possible position nodes.

Parameter m can be further tuned by using map information to distinguish between pedestrian-only areas and streets open to vehicular traffic [3]. This way, we obtain a map correlation of our first-level movements estimate. Using this approach we can substantially improve the tracking quality of our look-up table techniques. Our time-forwarding tracking technique takes full advantage of all available GIS information regarding area classification, so that validation of candidate location estimates depends on map constraints. In many cases, the additional information provided by the map is poor or absent; when this happens, our technique is able to dynamically build an information database by estimating all relevant knowledge including speed and acceleration.

4. Context Representation Issues

One of the most challenging issues faced by architectures supporting ambient-aware access control is the management of several dynamic aspects that define user context. Specifically, it is important to capture diverse sources of contextual information like the user's location, posture and expression. Several formats for describing mobile users' profile data have been proposed such as the *Presence Information Data Format* (PIDF) [19]. However, some context representation issues cannot easily be dealt with through user profiles. In particular, two major classes of problems have to be addressed: *i)* user context (i.e., digital identity) information should only be provided to authorized entities, and *ii)* solutions to ensure security and privacy of sensor data should be adopted. While there are several preliminary systems claiming to control access to information coming from sensor networks, a complete access control mechanism managing and relying on context data to protect access to

[3] A similar dynamic tuning of parameter m could be done based on the characteristics of the user's vehicle.

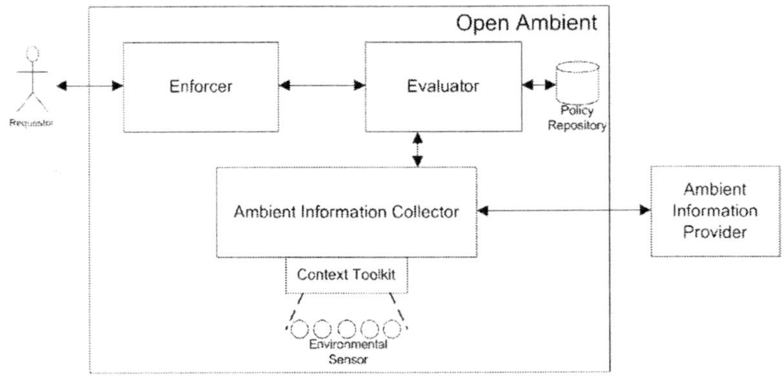

Figure 3. OpenAmbient Architecture.

resource is needed where sensor data are represented as first class citizens and described via suitably expressive metadata [8]. A major challenge in the representation of the context within ambient-based policies is the management of intrinsic uncertainty and error held by location and video measurement. In principle, such information could be represented by extending device oriented profiling languages such as *Composite Capabilities/Preference Profile* (CC/PP)[18]. Here, we adopt a different line of research, representing uncertainty underlying location and video sensor technologies in terms of a set of standard interfaces and semantically uniform *Service Level Agreement* (SLA) parameters negotiated between the involved parties to agree on the desired quality of service. The concept of Service Level Agreement (SLA) is used as the contractual means that the access control component, the ambient information collector and the end-user (for privacy purposes only) adopt to agree upon and set quality of services attributes to manage the intrinsic uncertainty held by context information.

5. Open Ambient Architecture

The OpenAmbient architecture is composed by four major components (see Figure 3) joining concepts coming from traditional access control architectures, and innovative idea for the management of context information lifecycle and communication protocols.

Enforcer. It represents the OpenAmbient external interface with the requestor. The Enforcer is the component responsible for the management and enforcement of the access control decision returned from the evaluator component. In particular, it grants to the users access to resources (e.g., an application service) only if a positive decision has been taken.

Evaluator. It represents the decision point. The evaluator is the component responsible for collecting the ambient-based policies and matching them against the requests submitted by the requestor. This component manages the intrinsic uncertainty of ambient-based predicates evaluation and it, finally, calculates the traditional *allow/deny* result.

Ambient Information Collector. It represents the component responsible for the interaction with the ambient data repository and/or technologies. It manages all the negotiation processes that generate the SLA representing an agreement on the ambient information reliability. The required reliability can vary depending on different scenarios in which the OpenAmbient architecture acts. The ambient information collector, then, contains also a Context Toolkit that acts as an ambient information provider, when the ambient information resides in the OpenAmbient boundaries, managing the information retrieval process. In this case, the environment technologies and sensors, gathering context information, are directly managed by the OpenAmbient architecture.

Ambient Information Provider. It represents the component able to retrieve the requested ambient information. In particular, the provider could be logically internal to the OpenAmbient (*Context Toolkit*) architecture in the case in which the information are provided by technologies directly managed by it such as camera. Otherwise, as in the case of mobile technologies, the provider could be an external service (e.g. OpenWave Location Manager [26]) that provides the requested data abstracting the underlying technologies.

OpenAmbient access control policies provide a high level of accountability and readability aimed at increasing user acceptance of high-secure infrastructures, assisted living and healthcare environments involving continuous monitoring and surveillance. From the functional point of view, OpenAmbient architecture faces the issues arising from the management of contextual information and uncertain measurements. First, it provides an infrastructure for the evaluation and enforcement of ambient-based policies. Then, OpenAmbient provides and implements different communication protocols managing the heterogeneity of the Ambient Information Provider. Finally, the OpenAmbient architecture supplies a general-purpose negotiation protocol used by the involved parties to agree upon a SLA contract that will drive any further interaction. In the remainder of this section we briefly discuss a running example illustrating security issues that can be addressed by the OpenAmbient approach.

Example 5.1 *As an example, we consider a highly-critical work environment, such as an automated bank. This environment includes services like an automatic* deposit drop box, *where users drop their deposits to be later collected by bank employees. Each drop box is protected by a timer that allows access only at pre-determined times. Also, access is granted only if the employee has been authenticated upon entering the box area by means of a key/badge. In this critical environment, however, an attacker could steal the badge of an authorized employee to access the drop box without the fulfillment of the requested conditions. Ambient-based conditions can prevent this type of attacks, making the (ubiquitous) access control infrastructure more secure and reliable. Suppose that the security administrator defines a policy allowing the access to the drop box only if the authorized bank employee is located inside the area where the box is according to reliable positioning techniques such as mobile phone positioning, RFID, GPS and so forth. In addition, the security administrator requires that the bank employee must be identified by the surveillance cameras as a member of a pool of authorized users, by means of face identification techniques. Finally, the administrator requires the bank employee to be alone in the area* *where the box is located. Syntactically, our policy has the following form.*[4]

```
bankEmployee WITH User.category = trustedUsers CAN access ON bankDepositBox IF
    density(bankDepositArea,1,1) AND faceIdentification(User,trustedPool) AND
                        inArea(User,bankDepositArea)
```

In particular, it is noticeable that every bankEmployee **User** *belonging to the* `trustedUsers` *category can access the deposit drop box only if she is identified by means of a face identification process and she is alone in the area where the box is located. A more complex system could state conditions involving a combination of subjects* **(User)**. *For instance, it could be required the trusted user to be accompanied by another authorized user. To this purpose, one could define the subject representing the access requestor as a conjunction of different subjects [27].*

6. Conclusion

Ubiquitous access control allows for writing and enforcing ambient security policies improving the overall security of critical workplaces or assisted living environments. The adoption of ambient-based conditions as a part of a sound access control model and architecture provides a set of functionalities improving the strength and robustness of ubiquitous access control process against malicious users.

Acknowledgments

This work was supported in part by the European Union within the PRIME Project in the FP6/IST Programme under contract IST-2002-507591 and by the Italian MIUR within the MAPS projects. This work was also partly funded by the Italian Ministry of Research under FIRB contract n. RBNE05FKZ2_004 TEKNE.

References

[1] US FCC enhanced 911 fact sheet 2001. http://www.fcc.gov/911/enhanced/release/-factsheet_requirements_012001.pdf, January 2004.

[2] Wölfle, G.; Hoppeand, R.; Zimmermann, D.; Landstorfer, F. M.: Enhanced Localization Technique within Urban and Indoor Environments based on Accurate and Fast Propagation Models. *European Wireless* 2002,Firence (Italy), February 2002.

[3] Wertz, P. ; Hoppe, R.; Zimmermann, D.; Wölfle, G.; Landstorfer, F. M.: Enhanced Localization Technique within Urban and Indoor Environments. *3rd COST 237 MCM-Meeting,* in Guildford, UK, January 2002.

[4] James, J.; Caffery Jr.; Stüber, G. L.: Overview of Radiolocation in CDMA Cellular Systems. *IEEE Communications Magazine*, April 1998.

[4]The language syntax has been defined in the context of European Project PRIME (Privacy and Identity Management for Europe, http://www.prime-project.eu.org/)

[5] Sharawi, M.S.; Aloi, D. N.: Investigation into performance of E-OTD for GSM users in Telematics applications. In *Proc. of conference at SPIE's 17th Annual AeroSense Symposium*, Florida - USA, April 2003.

[6] Anisetti, M.;Ardagna, C. A.;Bellandi, V.;Damiani, E.; Reale, S.: Positioning method and system for mobile communications networks, related network and computer program product. *European Patent* n.05425643.3, 2005.

[7] Anisetti, M.; Bellandi, V.; Damiani, E.; Reale, S.: Localization and tracking of mobile antenna in urban environment. In *Proc. of the IST 2005 - International Symposium on Telecommunications,* Shiraz, Iran, September 2005.

[8] Ardagna,C.A.; Cremonini,M.; Damiani, E.; De Capitani di Vimercati, S.; Samarati, P.: Supporting location-based conditions in access control policies. In *Proc. of the ACM Symposium on InformAtion, Computer and Communications Security(ASIACCS'06)*, Taipei, Taiwan, March 2006.

[9] Azzini, A.; Ceravolo, P.; Damiani, E.: Generating context metadata based on biometric systems. In *Proc. of the Knowledge-Based & Intelligent Information & Engineering Systems (KES)* 2005, Melbourne (Australia), September 2005.

[10] Bellandi, V.; Anisetti, M.; Beverina, F.: Upper-face expression features extraction system for video sequences. In *Proc. of the International Conference on Visualization, Imaging, and Image Processing (VIIP05)*, September 2005, 83-88.

[11] Bhal, P.; Padmanabhan, V. N: Radar: An in-building RF-based user location and tracking system. *IEEE INFOCOM,* 2000.

[12] Caffery, J. J.; Stüber, G. L.: Overview of radiolocation in cdma cellular systems. *IEEE Communications Magazine*, April 1998.

[13] Corallo, A.; Cremonini, M.; Damiani, E.; De Capitani di Vimercati, S.; Elia, G.; Samarati, P.: Security, Privacy, and Trust in Mobile Systems. *Mobile and Wireless Systems beyond 3G: managing new business opportunities*, Idea Group Inc, 2005.

[14] Covington, M. J.; Long, W.; Srinivasan, S.; Dey, A. K.; Abowd, G.D.; Ahamad, M.: Securing context-aware applications using environment roles. In *SACMAT* 2001, 2001.

[15] Damiani, E.; Anisetti, M.; Bellandi, V.; Beverina, F.: Facial identification problem: A tracking based approach. In *Proc. of the IEEE International Symposium on Signal-Image Technology and InternetBased Systems (IEEE SITIS'05)*, Yaoundé, Cameroon, November 2005.

[16] Damosso, E.: *Cost 231:digital mobile radio towards future generation systems*. Bruxelles, 1999.

[17] Ekman, P.; Friesen, W.: *Facial action coding system: A technique for the measurement of facial movement*. Consulting Psychologists Press, 1978.

[18] Franklin R.; Klyne, G.; Reynolds, F.; Woodrow, C.; Ohto, H.; Hjelm, J.; Butler, M. H.; Tran , L.: *Composite capability/preference profiles (cc/pp): Structure*. W3C Working Draft, 2000.

[19] Sugano, H.; Fujimoto, S.; Klyne, G.; Bateman,A. ; Carr,W.; Peterson, J.: *Presence information data format (PIDF)*. Network Working Group, IETF, Internet Draft, 2003.

[20] Geer, D.: *The E911 dilemma*. November/December 2001.

[21] Georghiades, A. S.; Belhumeur, P.N.; Kriegman, D.J.: From few to many: Illumination cone models for face recognition under variable lighting and pose. *IEEE Transactions on Pattern Analysis and Machine Intelligence,* 2001; Vol. 23(6), 643-660.

[22] Hallinan, P. W.: *A Deformable Model for the Recognition of Human Faces under Arbitrary Illumination*. PhD thesis, Harvard Univ. Cambridge, Mass., 1995.

[23] Hightower, J.; Borriello, G.: *Particle filters for location estimation in ubiquitous computing: A case study*. Nottingham, U.K., September 2004.

[24] Hoppe, R.; Wölfle, G.; Landstorfer, F.: *Fast 3-d ray tracing for the planning of microcells by intelligent preprocessing of the data base*. Paris, March 1999.

[25] Kalman, R. E.: A new approach to linear filtering and prediction problems. *Transactions of the ASME–Journal of Basic Engineering,* 1960; Vol. 82, 35-45.

[26] Openwave Location Manager: *http://www.openwave.com/*, 2006.

[27] Samarati, P.; Reiter M.K.; Jajodia S.: An authorization model for a public key management service. *ACM Trans. Inf. Syst. Secur.*, 2001; Vol. 4(4), 453–482.

[28] Schlit, B.: Ubiquitous location-aware computing and the place lab initiative. In *Proc. of the First ACM Intl. Workshop on Wireless Mobile Applications and Services on WLAN Hotspots,* San Diego, CA, September 2003.

[29] Sim, T.; Kanade, T.: Combining models and exemplars for face recognition: An illuminating example. In P*roc. of the CVPR 2001 Workshop on Models versus Exemplars in Computer Vision*, December 2001.

[30] Tian, Y.-L.; Kanade, T.; Cohn, J.: Dual-state parametric eye tracking. In *Proc. of the 4th IEEE International Conference on Automatic Face and Gesture Recognition (FG'00)*, Mar. 2000; 110-115.

[31] Turk, M. J.; Pentland, A.: Eigenfaces for recognition. In *IFIP Working Conference on Engineering for Human-Computer Interaction*, 1991; Vol. 3, 71-86.

[32] Vidal, J.; Najar, M.; Jativa, R.: High resolution time-of-arrival detection for wireless positioning systems. In *Proc of the IEEE 56th Vehicular Technology Conference,* 2002.

[33] Zhao, W.; Chellappa, R.; Phillips, P. J.; Rosenfeld, A.: Face recognition: A literature survey. *ACM Computing Surveys*, Dec. 2003; Vol. 35(4), 399-458.

[34] Zimmerman, D.; Baumann, J.; Layh, M.; Landstorfer, F.; Hoppe, R.; Wölfle, G.: Database correlection for positioning of mobile terminals in cellular networks using wave propagation models. *IEEE Vehicoular technology conference (VTC)*, Los Angeles USA, October 2004.

[35] Zimmerman, D.; Wertz, P.; Bauknecht, J.; Landstorfer, F.: *Performance of Location Methods in an Urban Mobile Communication Network*, Technical Report, 2004.

In: Long-Term and Dynamical Aspects of Information... ISBN 1-60021-912-8
Editors: Andreas U. Schmidt et al, pp. 73-85 © 2007 Nova Science Publishers, Inc.

Chapter 6

EVALUATION OF PC-BASED REAL-TIME WATERMARKING SYSTEM FOR LONG-TERM VIDEO PRESERVATION

Takaaki Yamada, [*] ***Isao Echizen,*** [†] ***Satoru Tezuka*** [‡]
Hitachi, Ltd., Kashimada, Kawasaki, 212-8567, Japan
Hiroshi Yoshiura [§]
The University of Electro-Communications
Chofugaoka, Chofu, 182-8585, Japan
Norihisa Komoda [¶]
Osaka University
Yamada-oka, Suita 565-0871, Japan

Abstract

New business opportunities of digital video such as IP-based inspection require for preserving the data during long term. Digital watermarks remain in the content and provide supplementary evidence of integrity. For them to become practical, a means is needed for processing them in real time on PCs. We previously proposed system architecture and a method to shorten watermarking time by reusing the estimated watermark imperceptibility among neighboring frames. They enable real time processing in a series of captured analog video signals: watermarking the video, encoding it into MPEG4 (QGVA, 1Mbps, 30fps), and storing the video for 12 hours in maximum. We have now evaluated the validity of the system as a software application.

1. Introduction

Digital video is being made widely available due to its advantages over analog content. It requires less space, is easier to process, and is not degraded by aging or repeated use.

[*]E-mail address: takaaki.yamada.tr@hitachi.com
[†]E-mail address: iechizen@sdl.hitachi.co.jp
[‡]E-mail address: tezuka@sdl.hitachi.co.jp
[§]E-mail address: yoshiura@hc.uec.ac.jp
[¶]E-mail address: komoda@ist.osaka-u.ac.jp

The wide availability of broadband networks and high-efficiency personal computers (PCs) enables live digital video content, such as live concerts and distance education, to be distributed over the Internet. Moreover, many digital video applications have been developed such as IP-based inspection. A serious problem, however, is that the copyright of digital content is easily violated because the content can be easily copied and redistributed through the Internet. Digital watermarking, which helps protect the copyright of digital content by embedding copyright information into it, is one countermeasure. Another problem is authenticating the content [1].

Original digital video data may be partially lost or become unreadable during long term preservation due to video codec changes and/or physical disk crashes. Digital watermarks can be used to establish the integrity of digital videos. They remain in the content itself for long time even if it is converted to another data format, and provide supplementary evidence relating copied content to the original content. Video watermark embedding in real time is an essential requirement for various emerging video applications. Systems have been developed for embedding video watermarks in real time that use dedicated hardware such as field programmable gate arrays (FPGAs) [2] and media processors [3]. Real-time embedding using hardware processing is well-suited to applications, such as broadcasting, in which video content must be distributed with high efficiency. However, hardware-based real-time embedding is problematic in terms of installation and maintenance cost and version upgrading.

A particular problem is preserving video for the long term. Therefore, systems are needed for which the functions can be easily expanded, that are easy to install, and that have a low initial cost. We have developed a real-time watermark embedding system that uses software and runs on an ordinary personal computer (PC) [4]. The proposed software-based system facilitates installation and version upgrades because it runs on the user's current computer. Additionally, the proposed system can interact with other systems and functions on the computer.

Section 2 presents a short overview of video watermarking and describes the previously-developed real-time watermarking system [4]. Section 3 describes its experimental evaluation and presents the results. Section 4 discusses the security aspects of digital video from the viewpoint of long term security. Section 5 briefly summarizes the key points.

2. System Outlook

2.1. Technical Requirements

From capture through distribution, conventional watermark (WM) processing is not suitable for real-time software applications because at each step the video data is stored in a buffer such as a hard disk. To achieve real-time processing, we combined all processes (capture, WM embed, and MPEG encode) so that the data are not stored in buffers on a hard disk but in cache and memory, and the time for WM embedding should be shortened to less than 1/30 second per frame (33ms).

Various WM embedding algorithms have been proposed [5, 6, 7]. We previously developed a video WM process [5] based on a basic patchwork algorithm [8]. Testing showed that the WMs it embeds by altering pixels have almost imperceptible alternations[5]. As

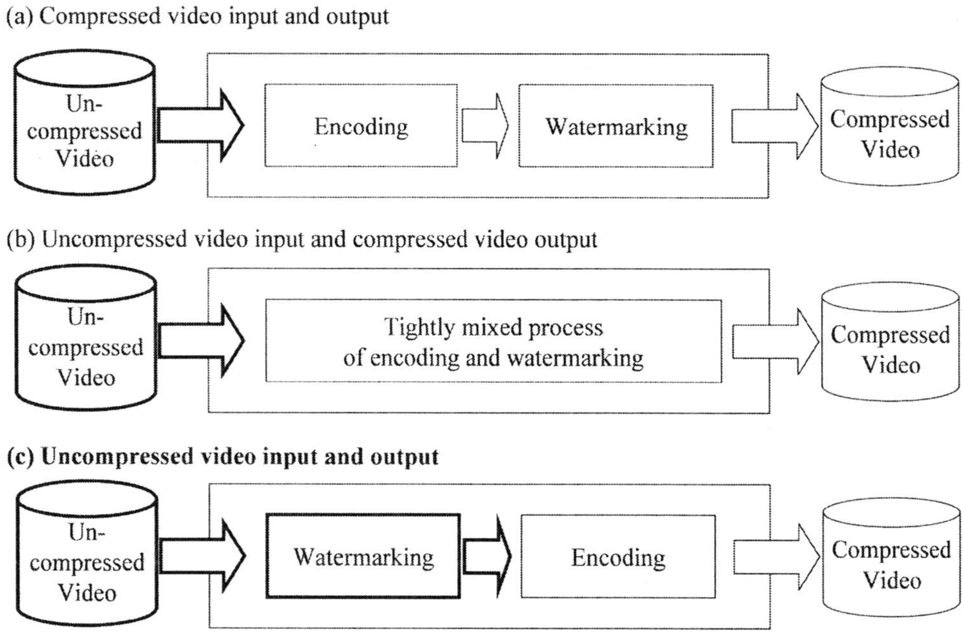

Figure 1. Three processes from the viewpoints of WM System I/O and WM algorithm I/O. Process (c), uncompressed video I/O for WM, is assumed.

shown in Figure 1, from the view point of system I/O, the uncompressed video input is compressed by an assumed system. However, both the input and output of our WM algorithm are uncompressed videos. Processing a huge data of uncompressed video require more resources than normally provided by a commodity PC, so real time WM embedding by software implementation has yet to be achieved. Our algorithm [5] is based on process (c) in Figure 1. Process (c) is advantageous because uncompressed video format is widely used as the input to many commercial video codec products. Based on the WM algorithm [5], we next briefly describe how our WM algorithm shortens the embedding time and thus enables real-time processing.

2.2. Improvement of Embedding Process

To shorten the watermark embedding process, we reduce the time needed for calculating the WM strength by reusing the estimated imperceptibility based on the properties of motion pictures; i.e., the properties of a frame are similar to those of the neighboring frames [4]. We reuse the WM strength of the previous frame as that of the current frame as follows. See feature analysis processes for calculating the WM strength in Figure 2. $s_i^{(t)}$ is the WM strength at pixel i on the t-th frame. If frame t, $(t = 0, 1, \ldots)$ satisfies $t \mod r = 0$, calculate its $s_i^{(t)}$, where $r \geq 2$ represents the interval over which the same WM strength is applied. Otherwise, reuse the $s_i^{(t)}$ of the last calculated frame:

$$s_i^{(t)} = s_i^{(t \mod r)} \quad \text{for all } i.$$

On the basis of the results of a preliminary analysis, we set r to 2 so that we achieved

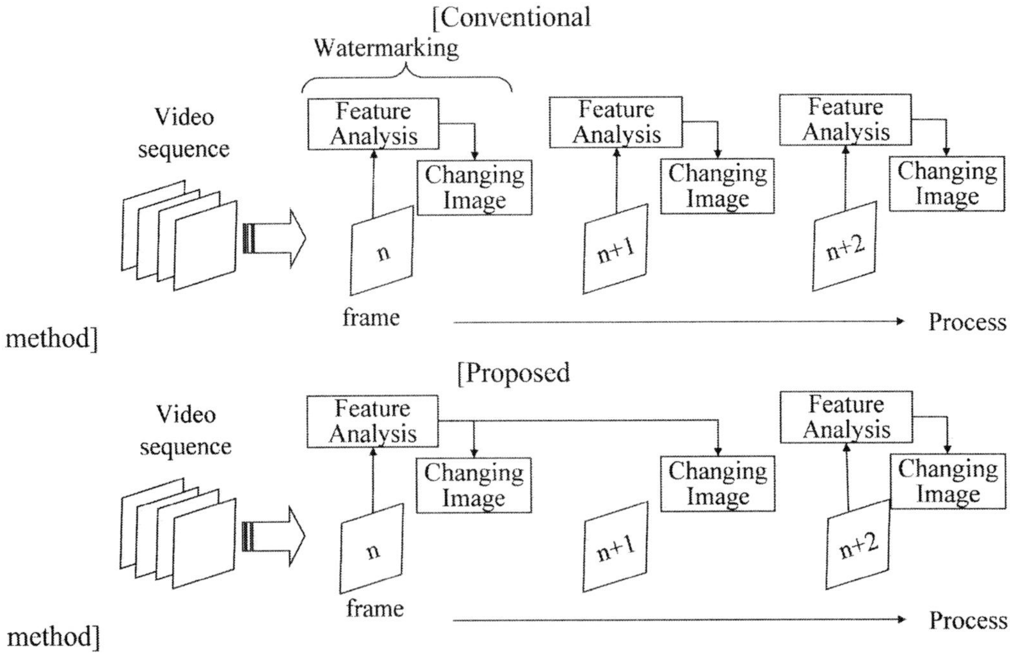

Figure 2. Reuse of WM strength.

real-time processing while maintaining picture quality. Figure 2(b) gives an overview of reusing the strength when $r = 2$.

2.3. Prototype

We developed a prototype real-time WM embedding system using a PC (Pentium[1] 4–3.4GHz) with a video capture board and an NTSC (National Television Standards Committee) video camera. Figure 3 shows the system architecture. Figure 4 shows the appearance of the system and the graphical user interface. The system specifications are shown in Table 1.

The system captures signals output from the video camera and embeds and stores the encoded data on the PC's hard disk. WMs representing 64 bits of information are embedded into the luminance signal (Y) of the video signal during MPEG-4 encoding, and the watermarked MPEG-4 bit stream is recorded on the hard disk. The embedding, encoding, and recording are all done in real time.

[1]Pentium is trademark or registered trademark of Intel Corporation or its subsidiaries in the United States and other countries.

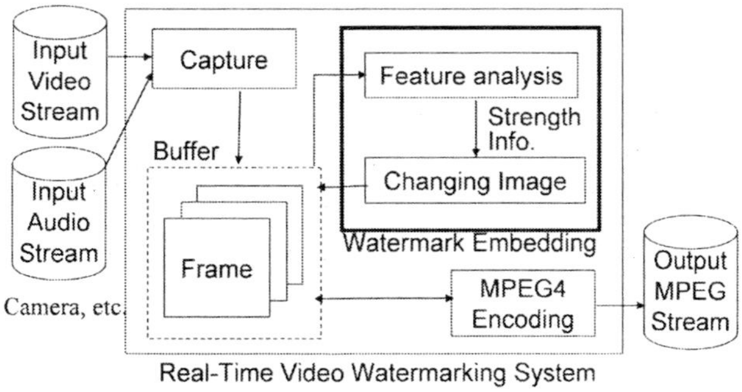

Figure 3. System architecture.

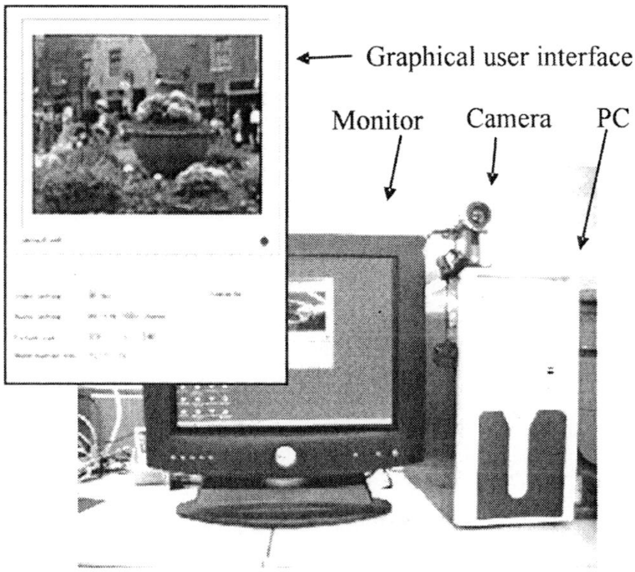

Figure 4. System outlook.

Table 1. System specifications.

PC	Ordinary PC with PCI bus CPU: Pentium 4 3.4GHz Memory: 1GB RAM
Video Input	NTSC RSA pin
Resolution	320 x 240 pixels (QVGA)
Frame rate	30 fps
Bit Rate	Video: 1024 kbps; Audio: 128 kbps
Payload	64 bits of information
Detection error rate	$\leq 10^{-10}$

3. Evaluation

3.1. Processing Time

3.1.1. Masurement Method

The prototype was evaluated using Standard video [9] and images with sounds taken using a video camera of a person walking in a room.
(1) Function and measurement items. The system consists of five functions: capturing images and sounds, watermarking, video encoding, audio encoding, and data storage. The capture is stably done using hardware without significant delay. Therefore, measuring the capture time is not required. The data storage, on the other hand, is problematic. Because the operation of the OS and the disk drive control disk cashes independent from the application software, it is essentially unpredictable when the buffer is flushed out. Therefore, the time measured for data storage is included in the time for video encoding. The time for audio encoding is independent of the other measurement. Continuous available time is calculated by error logs written out in case of dropping a frame. Five times were measured or calculated.

- Watermarking time
- Video encoding time
- Audio encoding time
- Total processing time (total of three times above)
- Continuous available time

(2) Additional operations. Since writing logs onto disk could cause disk access events that might effect our results due to insufficient PC resources, we wrote logs to memory and flushed them out after measurements. Unless memory was fully consumed by logging, the measure time was one second (30 frames). In addition, data in case of 20 seconds was also recorded for disk access events and synchronous delay events of audio.
(3) Recording. Any variances in the time to process a frame, 30 frames per a second, were recorded for the video encoding and watermarking. Since audio encoding time depended on the other sampling rate, 44.1KHz, we converted the measured data in proportion to the frame rate, 30 fps, as horizontal axis of result graphs below.

3.1.2. Masurement Results

(1) Watermarking time. Figure 5 shows the variance in the processing time for embedding watermarks using our shortened algorithm for standard video and perspective video taken in a room (the time with our algorithm is independent of the content). Skipping the process for feature analysis per two frames resulted in the peak time per two frames. This means that our proposed method shortens the mean watermarking time by about 63%. It takes 7ms versus 11ms with the conventional method, so it speeds up the process by 1.6 times.
(2) Total processing time. Disk access events are driven by the video encoding process. When compressed video data is stored at 1 Mbps, the transfer speed to the hard disk or to PC

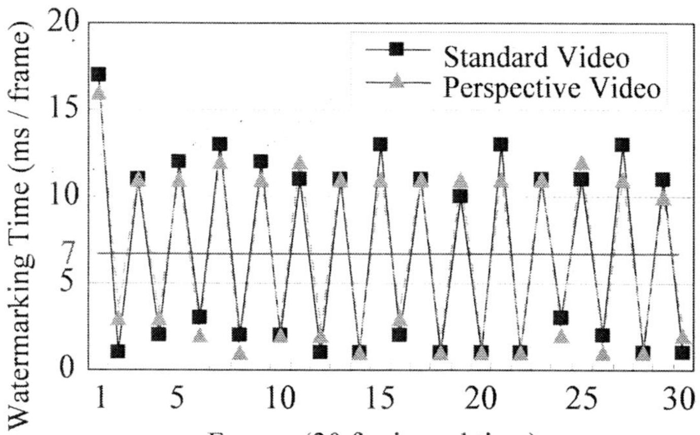

Figure 5. Watermarking time.

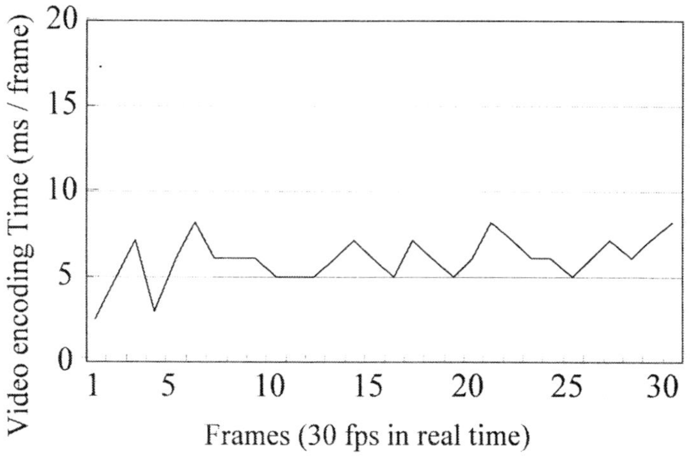

Figure 6. Video encoding time without disk accesses.

bus is not as critical to the total processing time as for uncompressed base-band video data. However, the video encoding process may temporally take longer to store partially encoded video on a hard disk. The OS controls disk access, and the disc access loads cannot be predicted. On the other hand, audio encoding can suffer synchronous delay due waiting for the results of video encoding. Therefore, we measured the total system processing time for both cases, meaning that the times for random events were included or not.

(a) Without disk accesses and synchronous delays. Figure 6 shows the variance in the video encoding time without disk access. Figure 7 shows the variance in the audio encoding time without synchronous delay. Since each process consumes a discrete amount of CPU power, the total system processing time of the system can be calculated by calculating their sum. The total processing time without disk accesses and synchronous delays was 16 ms.

(b) With disk accesses and synchronous delays. Figure 8 shows the variance in the video encoding time for the video of 20 seconds length. Figure 9 shows that of audio encoding.

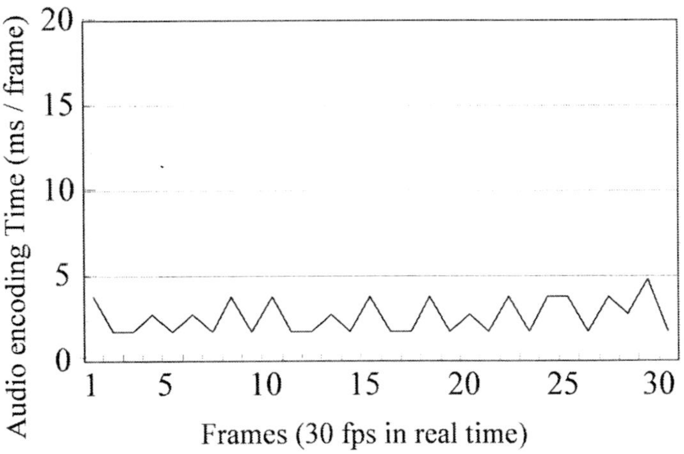

Figure 7. Audio encoding time without synchronous delays.

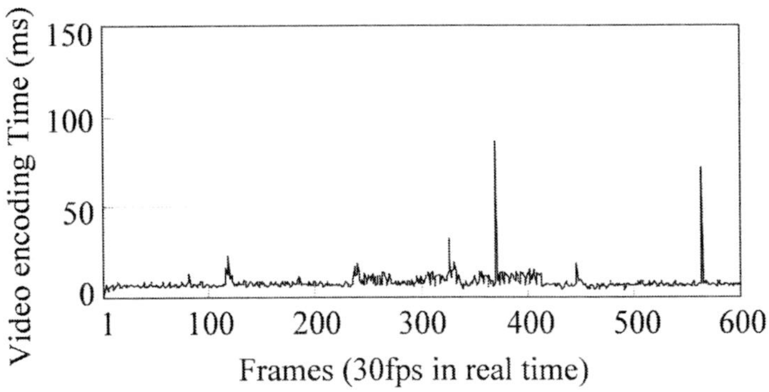

Figure 8. Video encoding time with disk accesses.

The results in both figures demonstrate that the system operated stably. The audio encoding process sometimes had to wait for video encoding process to finish. Since these processes are independent, it is difficult to distinguish delays caused by OS from the synchronous delays. The sum of the times shown in Figure 8, 9 and watermarking time might include idle time. See Figure 10.

The average total processing time with disk accesses and synchronous delays was 19 ms. Although delays often occurred due to disk accesses as shown in Figure 10, the temporal delays were absorbed in the buffers storing the video streams, so there were no frame drops. Moreover, no frames were dropped during 12-hours continuous operation experiment.

3.1.3. Summary of Processing Time Evaluation

Our proposed method speeded up the watermarking process by 1.6 times, enables a series of processes (capturing, watermarking, encoding and storing) to be stably performed at 19 ms per a frame for 12 hours. This is better than the 33 ms per frame given as a requirement.

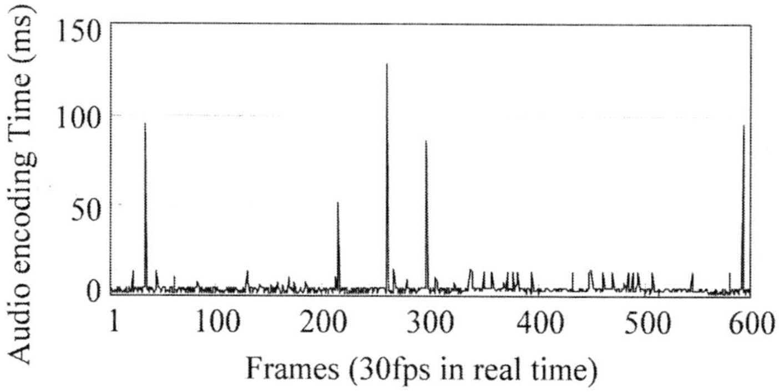

Figure 9. Audio encoding time with synchronous delays.

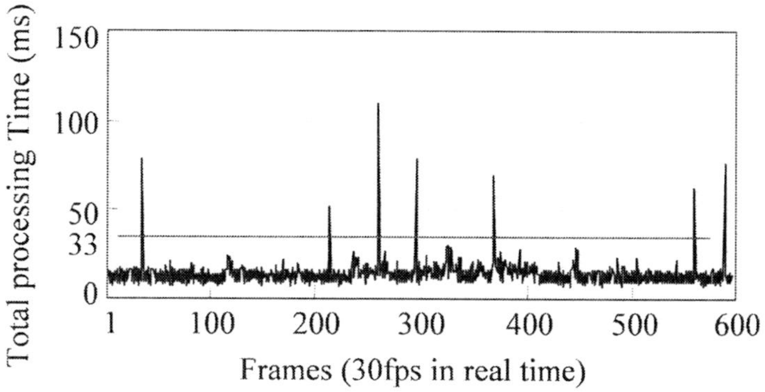

Figure 10. Total processing time.

3.2. Image Degradation

The image degradation of the watermarked image was evaluated using a standard method [10].

- Two watermarked images, SW1 and SW2, are generated from original image S with Parameter m is used to define how many times the results of feature analysis could be reused per frame. It is 1 for SW1 (representing the conventional method), and it is 2 for SW2 (representing the proposed method).
- Encoded image SE is made by compressing S under the condition of MPEG4, 30fps, and 1Mbps. SWE1 and SWE2 are made from SW1 and SW2 under the same condition.
- SW1 (SW2) is compared with SE using the subjective rating method.
  - STEP1: SW1 (SW2) is displayed to ten evaluators individually after displaying SE.
  - STEP2: The evaluators subjectively judjed the degree of degradation from SE to SW1 (SW2) using scores of 5,4,3,2 and 1 (imperceptible, perceptible but not annoying, slightly annoying, annoying and very annoying).

– STEP3: The average score for each image is used for the evaluation.

[EntranceHall] 

[WhaleShow] 

[WalkThroughTheSquare]

Figure 11. Sample pictures.

Three sample pictures are selected from standard videos, representing different degree of object movement: "EntranceHall" has little movement, "WhaleShow" has a moderate amount of movement, and "WalkThroughTheSquare" has a lot of movement. See Figure 11. As shown in Table 2, the evaluation results demonstrate that the proposed method performs the same performance as previous method.

Table 2. Evaluation of image degradation.

Sample pictures	Conventional method	Proposed method
EntranceHall	4.8	4.8
WhaleShow	4.6	4.6
WalkThroughTheSquare	4.6	4.6

A score of 4.5 represents the perceptible limit, the point at which the number of people who perceive the image degradation (score 5) equals the number of people who perceive some degradation but are not annoyed (score 4). The evaluated scores of 4.6 to 4.8 mean that the method works well enough for practical use. Theoretically, because the proposed method skips the feature analysis processes, it degrades image quality compared with the conventional method. However this degradation was apparently not perceptible.

3.3. Survivablity

Detection rate is defined as the proportion of frames in which a watermark is detected against all frames. Survivablity is evaluated on the basis of the detection rate using the same samples used to evaluate image degradation. The proposed method performed almost as well as the conventional method for all three samples. Further testing showed that the watermarks survived not only MPEG-4 encoding but also recompression, scaling, rotation,

Table 3. Security aspects of digital video.

Huge data files	The data management requires much time and space. Prone to operator error. Once watermarks are embedded, they do not only affect on the operator's tasks but also establish integrity of the data under the various operations.
Various types of metadata	Integrated metadata management is difficult. Operator errors may occur. Watermarks could combine metadata and content, so they are helpful for the metadata error detection.
Low readability	Migration or format conversion invalidates digital signatures. Signing again on the converted media requires continuous key management, which is costly over the long term. Remained watermarks in content prove the integrity of each version.

shifting, cutting out, and Gaussian filtering. Survivability was excellent in ten test cases using standard videos [9] having various motion properties.

3.4. Extensibility

Since the proposed method is implemented on a commodity PC, the system is flexibly extensible. For instance, the video codec can be easily exchanged from MPEG4 to another that uses the same computing power. If the PC hardware performance is improved, it can be switched to another video codec that requires more CPU power, such as the H.264 codec. Furthermore, if the encoded video file is transferred to a streaming server, the watermarking system can be extended into a live video streaming system.

4. Video Watermarks for Long-Term Security

Analog video is commonly accepted as evidence for legal matters because the original video is clearly different from a copied one. Digital video, in contrast, is looked at suspiciously. Better methods for authenticating digital video are needed, as mentioned above. However existing e-authentication technologies such as digital signature are not enough for long term security of digital video data, for the reasons summarized in Table 3.

Files containing digital video data are huge, so storage requires a vast amount of disk spaces. Moreover, various continuous operations are needed to load, move, backup, and restore the data. These are troublesome and are subject to operator error, meaning that authentication data can be easily invalidated. Metadata of digital video describes essential information such as date when the data was captured, and the person responsible for managing the data. There are various metadata standards, such as MPEG7, so operator error is problem in the integrated management of various types of video data. Readability of video data is low compared with digital still image data which has some standard formats for long-term preservation. In the near future, the recorded video format might change, for instance, from MPEG1 to MPEG2, MPEG4, or H.264. In addition, bit errors can occur in digitally recorded media. For example, the data storage lifetime on optical disks is estimated to be only five years in the worse case. Therefore, data migration is necessary for establishing readability. However, format conversion invalidates digital signatures.

Digital watermarks remain in the content over the long term, and provide supplementary evidence of content integrity. Emerging uses of digital video require digital watermarks to be implemented by software without dedicated hardware. The evaluation of our previously developed PC-based watermarking method demonstrated that it is suited for practical use.

5. Conclusion

Our real-time watermark embedding system is well suited for various video applications using software on an ordinary personal computer. Real-time processing is achieved by combining all the processes related to watermark embedding, thereby eliminating the need for storing huge amounts of uncompressed video data in buffers, and by shortening the watermark embedding time by reusing the estimated watermark imperceptibility among neighboring frames. Although the processing time is significantly reduced, the basic performance in terms of visibility and survivability are virtually the same as with the conventional method. Testing of a prototype system demonstrated the validity of this approach. Watermarks can help establish the integrity of video content preserved for a long time. This software-based watermarking system should thus find numerous applications.

References

[1] Cox, I. J.; Miller, M. L.; Bloom, J. A.: *Digital Watermarking*, Morgan Kaufmann Publishers, 2001.

[2] Owada, T.; Kitahara, J.; Yahata, T.; Echizen, I.; Fujii, Y.: Development of Hardware based Watermarking System, *SCIS2004*, 3D5-2, 2004. (in Japanese)

[3] Terada, K.; Fujikawa, Y.; Fujii, Y.; Echizen, I.; Yoshiura, H. ; Nishioka K.; Bandou, T.: Development of Real-time Video Watermarking System Using Media Processor, *Journal of the Institute of Image Information and Television Engineers*, 2004, vol. 58, no.12, pp. 1820-1827. (in Japanese)

[4] Echizen, I.; Yamada, T.; Fujii, Y.; Tezuka, S.; Yoshiura, H.: Real-time video watermark embedding system using software on personal computer, In: *Proc. of IEEE International Conference on Systems, Man and Cybernetics (SMC 2005)*, 2005, pp. 3369-3373.

[5] Echizen, I.; Yoshiura, H.; Taguchi, J.; Nakano, K.; Sasaki, R.: Maintaining Image Quality in Motion Picture Watermarking Based on Motion Detection, *IPSJ Journal*, 2002, vol. 43, no. 8, pp. 2529 - 2541. (in Japanese)

[6] Alattar, A. M.; Lin, E. T.; Celik, M. U.: DigitalWatermarking of Low Bit-Rate Advanced Simple Profile MPEG-4 Compressed Video, *IEEE Transactions on Circuits and Systems for Video Technology: Special Issue on Authentication*, Copyright Protection, and Information Hiding, 2003, vol. 13, no. 8, pp. 787-800.

[7] Hartung, F., Girod, B.: Watermarking of Un-compressed and Compressed Video, *Signal Processing*, 1998, vol. 66, no. 3, pp. 283-301.

[8] Bender et al.: Techniques for data hiding, In: *Proc. SPIE*, 1995, vol. 2020, pp. 2420-2440.

[9] The Institute of Image Information and Television Engineers, Evaluation video sample.

[10] Rec. ITU-R. BT. 500-7, "Methodology for the subjective assessment of the quality of television pictures", 1995.

In: Long-Term and Dynamical Aspects of Information... ISBN 1-60021-912-8
Editors: Andreas U. Schmidt et al, pp. 87-95 © 2007 Nova Science Publishers, Inc.

Chapter 7

LONG-TERM SECURE KEY ESTABLISHMENT

***Jens-Matthias Bohli,** *Jörn Müller-Quade*† *and Stefan Röhrich*‡**
Institut für Algorithmen und Kognitive Systeme / E.I.S.S.,
Universität Karlsruhe (TH), 76128 Karlsruhe, Germany

Abstract

In this paper we present a long-term secure key establishment protocol. Long-term security means resistance against attacks even if later, after completion of the protocol, some security assumptions become invalid. This makes the attacker more powerful, e.g. able to solve a formerly hard problem. In this setting we assume the hardness of the Diffie-Hellman problem during the protocol run, but later the attacker is able to solve the discrete logarithm problem which, e.g., is possible if quantum computers can be built. We achieve the security through combining the computational Diffie-Hellman key exchange with a classic Needham-Schroeder like protocol which uses symmetric encryption and a key distribution server. There are no known fundamental quantum attacks against good symmetric encryption schemes, which gives us long-term security. Through combination with the Diffie-Hellmann key exchange we achieve protection against a corrupted server as long as the Diffie-Hellman assumption holds and it allows us to give an efficient protocol with three rounds and five messages.

1. Introduction

Certain applications require confidentiality over a long period of time, e.g. medical records. To transfer such data between hospitals, encryption with a long-lasting security is needed. With symmetric encryption this seems to be possible—the usual requirement for the security of symmetric ciphers is that no better attack than brute force search is known and the cost of breaking the scheme will hence be close to exponential in the key length. However, for public-key cryptography the prediction of the cost of breaking the scheme is far more challenging, due to algorithmic advances in solving the underlying problem, which is usually much more structured as compared to symmetric ciphers. Especially today's

*E-mail address: bohli@ira.uka.de
†E-mail address: muellerq@ira.uka.de
‡E-mail address: sr@ira.uka.de

public-key schemes are threatened by the advent of quantum computers, which when built will efficiently calculate discrete logarithms and factor large numbers. In comparison, classical symmetric algorithms like AES aren't exposed to known efficient quantum attacks. Only a quadratic speed up for brute force attacks can be achieved by Grover's algorithm and it is believed that quantum computers cannot give an exponential speed-up for this kind of problem.

Key establishment protocols which are solely based on shared-key cryptography usually involve a trusted server for the key generation. A famous representative of this class of protocols is the shared-key protocol by Needham and Schroeder. However, a third party server may not be completely trusted, even if it is run by, e.g., the government. On the other hand, two-party key agreement protocols based on public-key cryptography offer the advantage that no third party must be trusted to that extent, as the session key will only be known to the participants. So, public-key cryptography has a benefit regarding confidentiality, whereas a trusted server can yield long-term security against adversaries like network providers who can access or store all the transmitted data for future cryptanalysis.

We present a practical protocol which combines the advantages of key establishment protocols authenticated by shared keys with a trusted server and public/private key pairs of the clients. The secrecy of the session key will depend on both, shared-key cryptography and public-key cryptography. Thus, even if the Diffie-Hellman (DH) assumption does no longer hold, the session key remains secret for the required period as long as the server does not reveal the long-term secret keys or ephemeral keys from the protocol runs. On the other hand, if the server is corrupted the protocol remains secure under the Diffie-Hellman assumption. Our protocol offers twofold protection without sacrificing efficiency.

2. Security Model

The protocol is analyzed in a security model close to [5] as a recent representative of the family of security models following [3, 4].

Participants. A set $\mathcal{U}$ of n users (including the server S) is given, where n can be polynomial in the security parameter. Each potential protocol participant $U_i \in \mathcal{U}$ may execute a polynomial number of protocol instances in parallel. Therefore each user U_i has *instances* $\Pi_i^{s_i}$, which may be taken for a process executed by U_i, that have assigned the seven variables $\mathsf{state}_i^{s_i}$, $\mathsf{sid}_i^{s_i}$, $\mathsf{pid}_i^{s_i}$, $\mathsf{sk}_i^{s_i}$, $\mathsf{term}_i^{s_i}$, $\mathsf{used}_i^{s_i}$ and $\mathsf{acc}_i^{s_i}$:

$\mathsf{used}_i^{s_i}$ indicates if this instance is involved in a protocol run;

$\mathsf{state}_i^{s_i}$ stores the state information during the protocol execution;

$\mathsf{term}_i^{s_i}$ indicates if the execution has terminated;

$\mathsf{sid}_i^{s_i}$ stores the session identifier that will serve as identifier for the session key $\mathsf{sk}_i^{s_i}$;

$\mathsf{pid}_i^{s_i}$ denotes the partner, i.e. the user who $\Pi_i^{s_i}$ aims at establishing a key with. To obtain a symmetric value $\mathsf{pid}_i^{s_i}$, we will include U_i himself in $\mathsf{pid}_i^{s_i}$;

$\mathsf{acc}_i^{s_i}$ indicates if the protocol instance was successful such that the session key is accepted;

$\mathsf{sk}_i^{s_i}$ stores the session key once it is accepted;

A more detailed description of the variables can be found in [2].

Initialization. Before the key establishment protocol can be executed, an initialization phase is needed where each user $U_i \in \mathscr{U}$ obtains a long-term secret key shared with the server S and a public key/private key pair (PK_i, SK_i). The keys are assumed to be generated honestly and the public keys are made available to all users and the adversary.

Communication network and Adversarial model. The network allows arbitrary point-to-point connections among the principals. However, the network is non-private and fully asynchronous. This is modelled by the adversary, who controls the network and delivers the messages. The adversary $\mathscr{A}$ is modelled as a probabilistic polynomially-time bounded (ppt) Turing machine who interacts with the instances via the following *oracles*:

$\mathsf{Send}(U_i, s_i, M)$ is used to send the message M to the instance $\Pi_i^{s_i}$. If $\Pi_i^{s_i}$ sends a message in the protocol as reply to M, the Send-oracle returns this message to the adversary. This oracle is also used to initialize a protocol run among principals U_A and U_B (and the server S). Therefore the adversary sends a special message $M = \{U_A, U_B\}$ with $U_A, U_B \in \mathscr{U}$ to an instance $\Pi_i^{s_i}$ with $\mathsf{used}_i^{s_i} = \mathsf{false}$. After such a query, $\Pi_i^{s_i}$ initializes its $\mathsf{pid}_i^{s_i}$-value to $\{U_A, U_B\}$, sets the $\mathsf{used}_i^{s_i}$-flag and processes the first step of the protocol.

$\mathsf{Reveal}(U_i, s_i)$ returns the session key $\mathsf{sk}_i^{s_i}$ if $\mathsf{acc}_i^{s_i} = \mathsf{true}$ and $\mathsf{sk}_i^{s_i} \neq \mathrm{NULL}$. Otherwise the distinguished NULL-value is returned.

$\mathsf{Corrupt}(U_i)$ returns both long term secret keys of U_i. Additionally to [4], we allow the adversary to use the Corrupt oracle to learn the secrets of the server S to be able to impersonate S.

$\mathsf{Test}(U_i, s_i)$ This query models the semantic security. The adversary $\mathscr{A}$ may use this query only once. If $\mathsf{sk}_i^{s_i}$ is defined and $\Pi_i^{s_i}$ has accepted, then with probability 1/2 the session key $\mathsf{sk}_i^{s_i}$ and with probability 1/2 a uniformly chosen random element from the according key space is returned. The adversary may continue other queries after the $\mathsf{Test}(U_i, s_i)$ call.

Correctness. To exclude "useless" protocols, a key establishment protocol is called *correct* if in the absence of active attacks a protocol run ends successfully, having all participants accepting a common key sk along with a common session identifier sid and a matching value pid..

Freshness. Now we have to define under which conditions the adversary should not know the session key and therefore not win the game induced by the Test-query. The instances holding such a key will be called fresh. Firstly we have to fix the following notion of partnering.

Definition 1. *Two instances* $\Pi_i^{s_i}$, $\Pi_j^{s_j}$ *are* partnered *if* $\mathsf{sid}_i^{s_i} = \mathsf{sid}_j^{s_j}$, $\mathsf{pid}_i^{s_i} = \mathsf{pid}_j^{s_j}$ *and* $\mathsf{acc}_i^{s_i} = \mathsf{acc}_j^{s_j} = \mathsf{true}$.

An instance $\Pi_i^{s_i}$ is called *fresh* if *none* of the following two conditions hold:

- For some $U_j \in \mathsf{pid}_i^{s_i}$ a $\mathsf{Corrupt}(U_j)$ query was executed before a query $\mathsf{Send}(U_k, s_k, *)$ with $U_k \in \mathsf{pid}_i^{s_i}$ has taken place.
- The adversary $\mathscr{A}$ queried $\mathsf{Reveal}(U_j, s_j)$ where $\Pi_i^{s_i}$ and $\Pi_j^{s_j}$ are partnered.

In these situations the adversary is allowed to know the session key, because he took part in the key exchange by corrupting a participant or learnt an outdated session key via Reveal, respectively.

Security. We consider only adversaries $\mathscr{A}$ who query Test once to a fresh instance and do not revoke the freshness of the Test-instance subsequently. The advantage $\mathsf{Adv}_{\mathscr{A}}(k)$ of such a ppt adversary $\mathscr{A}$ in attacking the protocol is defined as

$$\mathsf{Adv}_{\mathscr{A}} := |2 \cdot \mathsf{Succ} - 1|$$

where Succ is the probability that $\mathscr{A}$ outputs a correct guess for the secret bit of the Test-oracle.

We call the key establishment protocol P *secure* if for any ppt adversary $\mathscr{A}$ the function $\mathsf{Adv}_{\mathscr{A}} = \mathsf{Adv}_{\mathscr{A}}(k)$ is negligible. Furthermore we aim at key establishment protocols that fulfill the following two additional properties:

Definition 2 (Integrity). *A key establishment protocol fulfills* integrity, *if two honest instances accepting a protocol run with a matching session identifier* sid *hold with overwhelming probability identical session keys* sk *and agree about their respective partner, i.e. an identical value* pid.

Definition 3 (Entity-Authentication). Entity-Authentication *for an instance* $\Pi_i^{s_i}$ *is given,if* $\mathsf{acc}_i^{s_i} = \mathsf{true}$ *and, if the partner* $U_j \in \mathsf{pid}_i^{s_i}$ *is honest,* U_j *has with overwhelming probability an instance* $\Pi_j^{s_j}$ *with* $\mathsf{sid}_j^{s_j} = \mathsf{sid}_i^{s_i}$ *and* $\mathsf{pid}_i^{s_i} = \mathsf{pid}_j^{s_j}$.

3. Protocol

We present a protocol with five messages and three rounds, messages 1 and 2 respectively 3 and 4 can be sent in parallel. The protocol combines the modified STS-protocol [7, Protocol 5.16] with a substantial part of the Needham-Schroeder key exchange. The modified STS-protocol is basically identical to the Diffie-Hellman key exchange protocol in style of the ISO protocol [11], that was considered and proven secure in [8]. The parts of the STS-protocol are highlighted in bold face. For the Diffie-Hellman part of the protocol we assume a finite cyclic group with generator g, A and B choose their exponents a resp. b randomly. For authentication, public-key signatures from A and B are used, they are denoted by Sig_A for A's signature (resp. Sig_B for B's). The server shares symmetric encryption keys K_{AS} with A and K_{BS} with B, respectively. The encryptions are denoted by $\mathrm{Enc}_{K_{AS}}$ etc. The server generates a session key K_S which is used together with the Diffie-Hellman key to derive the final key.

1. $A \rightarrow S$: A, B, g^a

2. $A \rightarrow B$: $A, B, \mathbf{g^a}$

---

3. $S \rightarrow A$: $\mathrm{Enc}_{K_{AS}}(A, B, g^a, K_S), \mathrm{Enc}_{K_{BS}}(A, B, g^a, K_S)$

4. $B \rightarrow A$: $\mathbf{g^b}, \mathbf{Sig_B}(\mathbf{g^a}, \mathbf{g^b}, \mathbf{A})$

---

5. $A \rightarrow B$: $\mathrm{Enc}_{K_{BS}}(A, B, g^a, K_S)$, $\mathbf{Sig_A}(\mathbf{g^a}, \mathbf{g^b}, \mathbf{B})$

The final key K_{AB} is calculated through $K_{AB} = K_S \oplus \mathrm{KDF}(g^{ab})$ for a suitable key derivation function KDF, see e.g. [10]. The session identifier in the protocol is fixed as $H(g^a, g^b)$ for a collision resistant hash function H. The communication only runs between the initiating party A and the server S respectively between A and the partner B, no direct messages between S and B are needed, see Figure 1.

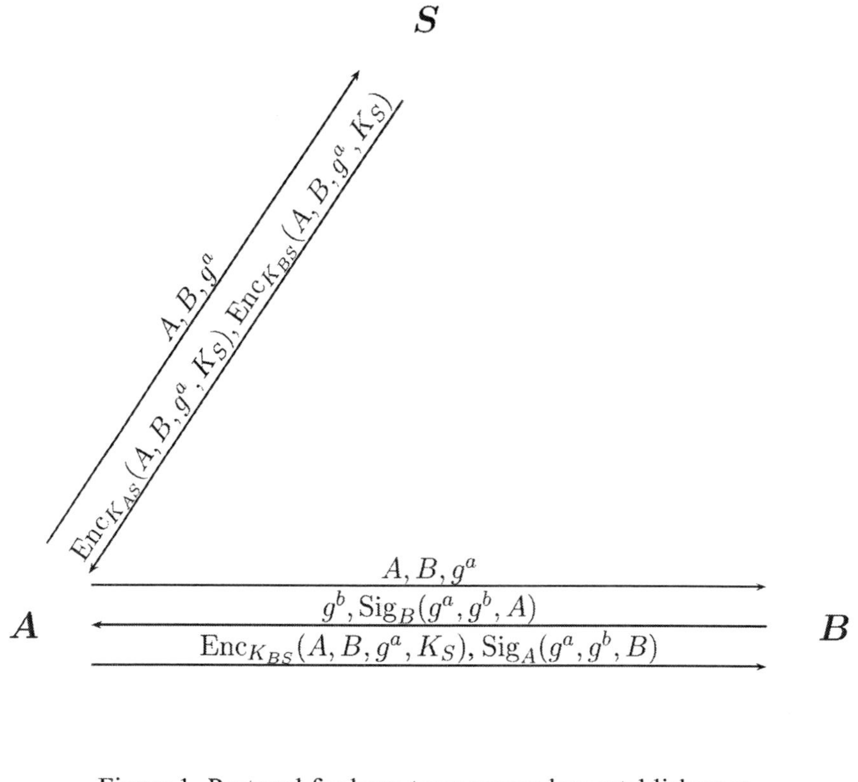

Figure 1. Protocol for long-term secure key establishment.

For similar reasons as Boyd's protocol [6] our protocol is more efficient as Gong's bound [9] of five messages and four rounds for an authentication only, nonce-based key exchange where both clients contribute to the key. This is achieved by assuring freshness through key contribution via the Diffie-Hellman part of the protocol which allows to save an extra round.

4. Security Analysis

The security of our protocol bases on the decision Diffie-Hellman assumption and the security of the symmetric encryption scheme. We express the security of the encryption scheme by the following game, that is close to left-or-right indistinguishability in [1], but involves simultaneous encryption under two keys.

Left-or-Right Experiment Let (Gen, Enc, Dec) be a symmetric encryption scheme. For two randomly chosen keys α and β encryption oracles $\mathsf{left}(l,r)$ and $\mathsf{right}(l,r)$ are given that return $(\mathrm{Enc}_\alpha(l), \mathrm{Enc}_\beta(l))$ and $(\mathrm{Enc}_\alpha(r), \mathrm{Enc}_\beta(r))$, respectively. The encryption scheme is said to be secure, if any ppt adversary $\mathscr{A}$ has an at most negligible advantage

$$\mathsf{Adv}_{\mathscr{A}} := \Pr\left[\alpha,\beta \leftarrow \mathrm{Gen} : \mathscr{A}^{(\mathsf{left}(\cdot;\cdot))} = 1\right] - \Pr\left[\alpha,\beta \leftarrow \mathrm{Gen} : \mathscr{A}^{(\mathsf{right}(\cdot;\cdot))} = 1\right].$$

We further assume, that the security of the encryption scheme holds independently of the Diffie-Hellman assumption or the hardness of computing discrete logarithms in finite cyclic groups. This assumption seems to be justified for common symmetric encryption schemes like, say, AES.

Proposition 1. *We assume that the symmetric encryption scheme* (Gen, Enc, Dec) *is secure in the above-mentioned sense of left-or-right indistinguishability. The key exchange protocol described above is* long-term secure *in the sense that it is a secure key establishment protocol if at least one of following conditions holds:*

1. *the server is uncorrupted,*
2. *the Diffie-Hellman assumption holds.*

If during the protocol runtime the Diffie-Hellman assumption holds and the server is uncorrupted, entity authentication and integrity are guaranteed by the protocol.

Proof. At first, two events are defined. Forge denotes the event that the adversary succeeds in forging an authenticated message (message 4 or 5) without having queried $\mathsf{Corrupt}(U_i)$ for the respective sender. An adversary $\mathscr{A}$ that can provoke Forge can be transformed in a forger for the signature scheme: A given public key is assigned to one of the n principals, $\mathscr{A}$ produces a forgery for the challenge key with a probability of $\geq \frac{1}{n} \cdot P(\mathsf{Forge})$. For an existentially unforgeable signature scheme S, the probability of Forge is bounded by

$$P(\mathsf{Forge}) \quad \leq \quad n \cdot \mathsf{Adv}_S^{\mathrm{cma}},$$

where $\mathsf{Adv}_S^{\mathrm{cma}}$ denotes the advantage of the adversary in violating the existential unforgeability under adaptive chosen message attacks.

The second event Repeat denotes the situation where a nonce N_i, a server key K_S or an ephemeral Diffie-Hellman key g^i, respectively, is used twice by any two instances. The values are only produced in response to a Send query. Let q_s be the polynomial bound for the number of queries to the Send oracle, then the probability of Repeat is bounded by

$$P(\mathsf{Repeat}) \leq \frac{q_s^2}{2^k}.$$

In case the adversary can bring about an event Forge or Repeat we declare the experiment as successful for the adversary. In the following we assume that neither event occurred and are interested in the adversary's success probability for this case.

Diffie-Hellman assumption holds. If the Diffie-Hellman assumption holds, the modified STS-protocol is secure (cf. [8]). We build the proof of our protocol on the security of the modified STS-protocol by reducing an attack to our protocol to an attack on the modified STS-protocol.

Let $\mathcal{A}$ be an adversary to our protocol. We construct an adversary $\mathcal{M}$ to the modified STS-protocol as follows: $\mathcal{M}$ is connected to a model for the modified STS-protocol. To simulate a model for our protocol, $\mathcal{M}$ will simulate user instances which are in a one-to-one relation to the instances in the STS-model. Additionally, $\mathcal{M}$ simulates the instances of the server and begins to initialize the users with symmetric keys they share with the server. Then $\mathcal{M}$ is able to simulate all oracle queries for $\mathcal{A}$:

Send(U_i, s_i, M): On a test query that initiates a session, $\mathcal{M}$ obtains g^a from the STS-model by initiating a session, and returns messages for the rounds one and two to $\mathcal{A}$. If M is a message of the first round, and U_i is the server, $\mathcal{M}$ will create a key K_S and return a message for Round 3. If M is a message for the second round and U_i is a user, called B in the protocol, $\mathcal{M}$ will obtain a value g^b and the signature $\mathrm{Sig}_B(g^a, g^b, A)$ from a Send query to the STS-model, and produce a Round 4 message. If M is a message of the third round, $\mathcal{M}$ will decrypt the message, check if it contains A, B, a valid key K_S and the value g^a sent in the first message. For a message M of the fourth round, $\mathcal{M}$ will forward the message to the according instance in the STS-model to obtain the signature $\mathrm{Sig}_A(g^a, g^b, B)$. Once both messages of rounds three and four arrived, $\mathcal{M}$ sends a message for Round 5. The instance will store the key K_S and accept.

Reveal(U_i, s_i): If the instance $\Pi_i^{s_i}$ has accepted and stores a value K_S, $\mathcal{M}$ will query Reveal to the according instance in the STS-model, yielding a key $\mathrm{KDF}(g^{ab})$, and return $K_{AB} = K_S \oplus \mathrm{KDF}(g^{ab})$.

Test(U_i, s_i): If the instance $\Pi_i^{s_i}$ has accepted and stores a value K_S, $\mathcal{M}$ will query Test to the according instance in the STS-model, yielding a key K, and return $K_{AB} = K_S \oplus K$.

The oracle Corrupt is answered in a straightforward manner. Note that Corrupt is allowed to be addressed to the server, returning all symmetric keys shared between the server and any user.

Honest server. We assume that the server is honest, i.e. the adversary is not allowed to query Corrupt(S) in order to learn all symmetric long-term keys. However, the adversary may break the Diffie-Hellman assumption or even be able to compute discrete logarithms after the last protocol execution.

We construct an Adversary $\mathcal{D}$ who can break the left-or-right indistinguishability of the encryption scheme (Gen, Enc, Dec) using $\mathcal{A}$ as a black box. Therefore, $\mathcal{D}$ will simulate the instances and oracles for the adversary. $\mathcal{D}$ will receive a challenge for the encryption scheme with Enc- and Dec-oracles for two keys. These oracles are assigned to two randomly chosen users and $\mathscr{D}$ generates symmetric keys for the remaining users. $\mathscr{D}$ chooses an index $i \in \{1,\dots,q_s\}$ and will for the ith situation when $\mathscr{D}$ has to chose a key K_S for a protocol run between the users A and B ask the left-or-right challenge $((A,B,g^a,K_1),(A,B,g^a,K_2))$ for the encryption scheme and use the returned ciphertexts as message in the protocol. If $\mathscr{A}'$ uses this session successfully as Test session, $\mathscr{D}$ will successfully answer the left-or-right challenge.

Putting everything together we obtain for the valid Diffie-Hellman assumption

$$\mathsf{Adv}(\mathscr{A})^{DH} \leq \mathsf{Adv}_{\mathsf{STS}} + P(\mathsf{Forge}) + P(\mathsf{Repeat}),$$

and for an honest server

$$\mathsf{Adv}(\mathscr{A})^{S} \leq n^2 q_s \mathsf{Adv}_{\mathsf{Enc}} + P(\mathsf{Forge}) + P(\mathsf{Repeat}).$$

Here, $\mathsf{Adv}_{\mathsf{Enc}}$ and $\mathsf{Adv}_{\mathsf{STS}}$ are the advantages of a ppt adversary against the symmetric encryption and the modified STS-protocol, respectively.

Integrity and entity authentication. These are runtime properties and we assume the server to be uncorrupted and the Diffie-Hellman assumption to hold.

Entity authentication is provided: the session identifier $\mathsf{SID} = \mathsf{H}(\mathsf{g}^{\mathsf{a}},\mathsf{g}^{\mathsf{b}})$ is computed from g^a and g^b. An instance that accepts a session has received a signed message including g^a and g^b from its intended partner. The partner instance knows the same values and must hold the same session identifier.

Also integrity holds: Two instances agreeing on the session identifier $\mathsf{SID} = \mathsf{H}(\mathsf{g}^{\mathsf{a}},\mathsf{g}^{\mathsf{b}})$ agree on g^a, g^b and therefore compute identical keys $\mathrm{KDF}(g^{ab})$. An honest server puts identical keys K_S in the encryptions that are both liked to the fresh nonce g^a. Thus, both instances compute the same key $\mathsf{sk} = K_{AB}$.

□

5. Conclusions

We presented a key agreement protocol that offers strong security guarantees. The protocol combines authentication by signature keys and shared keys with a trusted server. Security in the sense of implicit key authentication is provided as long as one assumption holds: Either the server remains uncorrupted or the Diffie-Hellman assumption holds. This results in a long-term security, particularly if the server keys are regularly changed. During runtime, if both assumptions are valid, the protocol offers entity authentication and integrity.

References

[1] Bellare, M.; Desai, A.; Jokipii, E.; Rogaway, P.: A Concrete Security Treatment of Symmetric Encryption. In *Proceedings of the 38th Annual Symposium on Foundations of Computer Science (FOCS'97)*, 1997, pp 394–403.

[2] Bellare, M.;Pointcheval, D.; Rogaway, P.: Authenticated Key Exchange Secure Against Dictionary Attacks. In Advances in Cryptology – EUROCRYPT 2000; Preneel, B., Ed.; *Lecture Notes in Computer Science* **1807**, Springer-Verlag, 2000, pp 139–155.

[3] Bellare, M.; Rogaway, P.: Entity Authentication and Key Distribution. In Stinson, D. R., Ed.; Advances in Cryptology – CRYPTO '93, *Lecture Notes in Computer Science* **773**, Springer-Verlag, 1993, pp. 232–249.

[4] Bellare, M.; Rogaway, P.: Provably Secure Session Key Distribution: The Three Party Case. In *Twenty-Seventh Annual ACM Symposium on Theory of Computing, Proceedings of STOC 1995*, ACM Press, 1995, pp. 57–66. Full version online available at http://www.cs.ucsd.edu/users/mihir/papers/3pkd.ps.

[5] Bohli, J.-M.; González Vasco, M. I.; Steinwandt, R.: Secure Group Key Establishment Revisited. *International Journal of Information Security*, to appear. preliminary version: http://eprint.iacr.org/2005/395/.

[6] Boyd, C.: A Class of Flexible and Efficient Key Management Protocols. In: *Computer Security Foundations Workshop*, 1996, IEEE, 1996, pp. 2–8.

[7] Boyd, C.; Mathuria, A.: *Protocols for Authentication and Key Establishment.* Springer, 2004.

[8] Canetti, R.; Krawczyk, H.: Analysis of Key-Exchange Protocols and Their Use for Building Secure Channels. In Advances in Cryptology – EUROCRYPT 2001; Pfitzmann, B., Ed.; *Lecture Notes in Computer Science* **2045**, Springer-Verlag, 2001, pp. 453–474.

[9] Gong, L.: Efficient Network Authentication Protocols: Lower Bounds and Optimal Implementations. *Distributed Computing*, **9**(3):131–145, 1995.

[10] *IEEE P1363-2000, Standard Specifications for Public Key Cryptography*, 2000.

[11] ISO/IEC IS 9798-3, "Entity authentication mechanisms – Part 3: Entity authentication using asymmetric techniques", 1993.

In: Long-Term and Dynamical Aspects of Information... ISBN 1-60021-912-8
Editors: Andreas U. Schmidt et al, pp. 97-111 © 2007 Nova Science Publishers, Inc.

Chapter 8

Applicability of Quantum Cryptography for Securing Mobile Communication Networks

Michael Marhöfer
Siemens AG, Siemens Networks GmbH & Co. KG
P.O Box 80 17 60, D-81617 Munich, Germany
Ilse Wimberger
Siemens AG Österreich, Program and System Engineering
Siemensstraße 90-92, A-1211 Vienna, Austria
Andreas Poppe
Faculty of Phyiscs, University of Vienna
Boltzmanng. 5, A-1090 Vienna, Austria

Abstract

After 20 years of basic research, quantum cryptography has meanwhile led to first commercial products. Its progress has triggered high publicity and additional R& D funding. The aim of this paper is to evaluate the status of quantum cryptography regarding its practical applicability for securing (mobile) communication networks. With that aim in mind, the paper presents a survey of the state-of-the-art, an analysis of its practical constraints and still open R& D challenges, and some candidate applications for securing mobile communication networks. First industrial applications of quantum cryptography have become reality; more applications may soon reach practical maturity due to recent technological progress.

1. Introduction

Since nearly 100 years, quantum physics continues to deliver its astonishing findings. Recently, some books have been published to introduce also the laymen to that field and its applications. Over the last 20 years, quantum cryptography [1, 2, 3] has established itself as a promising application of quantum physics [4, 5, 6].

The aim of this paper is to evaluate the status of quantum cryptography regarding its practical applicability for securing (mobile) communication networks. With that aim in mind, the paper presents a survey of the state-of-the-art, an analysis of its practical constraints and still open R& D challenges. It suggests some candidate applications of quantum cryptography for securing the operations of mobile communication networks. While research like in the EU-funded SECOQC project continues to improve quantum cryptography and to extend its reach [7], the promise of unconditionally secure key exchange has already led to a first series of commercial products and attracted early adopters, who now are evaluating this new technology under the specific constraints of their industry. In a time, when even quantum computing is no longer science fiction, quantum cryptography is now reaching industrial applications.

2. Practical Maturity of Quantum Cryptography

During the last years many experiments in the field of quantum cryptography were reported in scientific papers and triggered popular articles afterwards. But what is quantum cryptography? First of all, it is the name of the research discipline that tries to apply the principles of quantum physics for cryptographic purposes. Among the main principles which are used for quantum cryptography are the following:

- It is not possible to determine simultaneously the position and the momentum of a particle with arbitrary high accuracy (Heisenberg's uncertainty principle).
- It is not possible to measure the polarization of a single photon in the vertical-horizontal basis and simultaneously in the diagonal basis.
- Each measurement of the quantum state modifies the quantum state.
- It is not possible to copy quantum states (no-cloning-theorem).
- The duality of particles and waves of single particles culminate in hardly understandable effects without classical analog (e.g. superposition, entanglement).

In 1984 Bennet and Brassard published the first quantum cryptographic communication protocol, called BB84 [1], proposing quantum key distribution (QKD). QKD is a methodology for generating and distributing random encryption keys using the principles of quantum physics. The keys are secure from eavesdropping or tampering during transit. QKD ensures that the knowledge of any third party of the key is reduced to the level of guessing.

The key can be used as a one-time pad cipher to encode a message. The combination of quantum key distribution with a one-time pad cipher and an information theoretically secure message authentication scheme is referred to as quantum cryptography (QC). In this sense, quantum cryptography is unconditional secure which means that it provides an encryption that no analysis can break, irrespective of whatever advances are made in mathematics or computer science including quantum computation. With today's technology it is not possible to use quantum cryptography as a one-time pad stream cipher encoding typical data traffic, because the key generation rates achieved with QKD are too low by many orders of magnitude. Therefore hybrid systems are used, i.e. keys are generated and distributed with QKD and these keys are used in conventional encryption algorithms like e.g. AES.

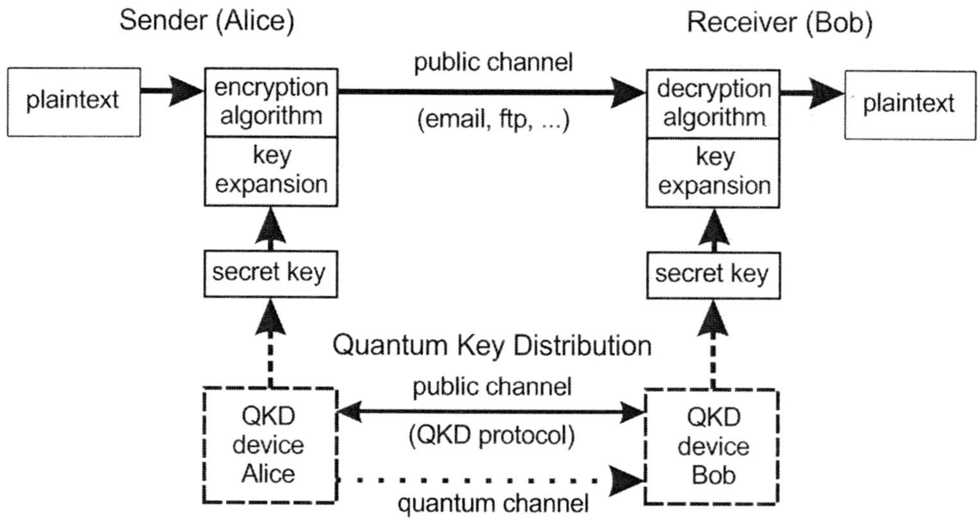

Figure 1. Hybrid quantum cryptography system. The QKD system (QKD device Alice and Bob are connected by a quantum and a public channel) builds up and refreshes a shrunk secret key that can be expanded to encrypt a plaintext at the sender and decrypt it at the receiver side.

In QKD systems the information is carried by the polarization or phase of single photons. The photons are sent over an optical fiber or free-space. At the receiver's site the photons are detected. These measurements build the so-called "raw key". Only if the randomly chosen quantum state preparation at the sender matches the settings for the basis at the receiver, both communication parties will generate the same bit value for sure in an ideal system. Therefore the basis of the measurements, but not the result is publicly announced in the procedure called "sifting". If no eavesdropping has happened, the sifted key bits will be perfectly correlated. Any action of an eavesdropper results in an increased error rate. Also in practice the sifted key contains errors due to background photons, detector noise and polarization imperfections. Even these errors must be attributed to a potential eavesdropper. Subsequently different algorithms for detecting and correcting these errors are applied. All the potential information that leaked out of the system due to imperfection or real eavesdropping is removed by the process called "privacy amplification". The result of this process is a secret key which contains much less bits than the original raw key. This is the price to pay for the detection of an eavesdropper in QKD.

For potential users the most important characteristics of a QKD system are the final key generation rate (= number of secret bits per second) and the transmission distance (= distance between key generator and detector). Ideally, the information is carried by single photons and a loss in the transmission line results in a decreased key generation rate. Therefore the key generation rate is strongly dependent on the distance and also varies a lot for the different QKD technologies available nowadays.

The optimal source for QKD produces only one photon (and not a second one). Quantum dots in semiconductors would theoretically do this job best, but are far from technical realization, even in the lab. Therefore QKD must use some workarounds: Today the most mature QKD technology is the technology based on weak coherent pulses (WCP). If pulses from a laser are attenuated to the single photon level, the photons are not equally distributed over the pulse train. To prevent all feasible attacks that are possible if more than one photon carrying the same information is in a pulse, a very low average number of photons per pulse and a corresponding low rate would be necessary. On the other hand, the special strength of WCP is that conventional diode lasers and standard single-mode optical fibers can be used. First commercial systems based on this technology are available (see chapter 2.1).

Decoy-states QKD [8, 9] are a counter-strategy to probe the most challenging attack of a possible eavesdropper. Some more ways have been developed to limit the adversary (e.g. additional coherence between states). These new protocols are currently subject of intense investigations by many research groups around the world [10].

High security is provided by QKD systems using entangled photons [11], because the key does not exist until the photons are measured. First demonstrations of QKD with entangled photons using standard telecommunication optical fibers up to 100km are on the way [12].

QKD based on continuous variables [13] uses polarized coherent states to generate the key, which leads to high secret bit rates for short distances. Standard components and optical fibers can be used, but security is not guaranteed when using lossy transmission systems. Reverse reconciliation is currently proven to overcome this maximal 3dB-loss-limit [14].

The optical link for transferring the photons in QKD systems may be either an optical fiber or just the atmosphere. Non-telecom-standard single-mode fibers with wavelength around 800nm are used in experiments, because efficient detectors are commercially available. The drawback is the high attenuation, so that these systems can only be used for short distances (below 5km). The ability of telecommunication optical fibers with wavelength near 1300 or 1550nm for long transmission distances is also perfect for quantum signals, but only poor single photon detectors are available for those wavelengths.

2.1. Products for QKD

The US company MagiQ Technologies [15] was founded in 1999. Since 2003 it has offered the MagiQ QPN Security Gateway, which is a VPN security solution based on quantum cryptography. In November, 2005 a new pricing for its QPN 7505 quantum encryption solution and the completion of several important field tests for the device was announced. The features for the QPN 7505 include:

- Long haul network support, including the ability to cascade two systems to reach up to 160km secure distance.
- Gigabit data channel integrated with quantum keys.
- Remote monitoring and management.
- Multiplexing of data and quantum channel over a single fiber, reducing the cost of deployment The price for a typical installation of the QPN 7505 starts at $97.000.

The Swiss company idQuantique [16] was founded in 2001 and is a spin-off of the University of Geneva. Besides the plug-and-play quantum key distribution system Clavis, which is designed for research and development applications, it offers the following commercial products:

- Vectis Link Encryptors are hardware quantum cryptography appliances for point-to-point wire-speed link encryption. Quantum key distribution and AES encryption engines are combined in a stand-alone unit. Vectis is a layer 2 encryption device that securely bridges two Fast Ethernet fiber optic networks. Keys are exchanged at regular intervals, up to 100 times per second. Vectis uses a patented auto-compensating optical platform to secure key exchange over distances of up to 100 km. Central monitoring and management of distributed Vectis link encryptors is possible within an enterprise network.

- Quantis is a quantum random number generator. The product is sold in two configurations (a PCI card and an OEM module) and can be easily integrated in existing applications.

The French company SmartQuantum [17] was founded in 2004. In April 2005 it announced the first complete commercial quantum security solution, called SQBox, which combines quantum key distribution, intrusion detection, high bandwidth encryption and monitoring technologies. SQBox has the following features:

- Maximum point-to-point distance: 80km.

- Physical level encryption with AES 192 bit by default (ability to support any encryption algorithm).

- Seamless network integration.

2.2. Overview on Recent Experimental Results

Between 2003 and 2006 roughly the same amount of experiments was published as between 1990 and 2002. Due to the increased relevance of the most recent results, only a few highlights are summarized here.

Experiments with Entangled Photons. In April 2004 an internet bank transfer was demonstrated in Vienna between the Vienna City Hall and the headquarters of a national bank. The transfer was secured with a key that had been retrieved by quantum key distribution based on entangled photons, which had been sent over a 1.45 km special single-mode optical fiber for 810 nm [18]. In December 2004 the same research group presented a prototype for secure communication (VPN tunnel) [19]. Since 2005 the Viennese researchers worked on enabling their QKD system for standard telecommunication fibers (1550 nm wave length) and subsequently so distances up to 100km [12] and key generation rates more than 1300 secure bit/second could be achieved over 25km [20].

Free-space Experiments. In 2005 it was reported that Chinese scientists succeeded in an inner-city free-space distribution of entangled photon pairs over a distance of 10.5km [21]. Nearly at the same time, an experiment was published where entangled photons were distributed directly through the atmosphere to a receiver station 7.8 km away over the city of Vienna [22]. Depending on atmospheric conditions, the amount of air between a base station and a satellite in low earth orbit is equivalent to 5 to 10 kilometers of air at ground level. Furthermore, in 2006 an experiment was published that realized a free-space distribution of entanglement and single photons over 144 km [23] between two islands. The range achieved thereby demonstrates the feasibility of quantum communication in space, involving satellites or the International Space Station (ISS).

Improvements in Distance, Key Generation Rate or System Availability. In 2004 Toshiba UK reported the first demonstration of quantum key distribution over a standard telecom fiber exceeding 100 km in length. With further improvements the system could be extended to 165km [24]. One year later Toshiba UK demonstrated a robust, compact and automated quantum key distribution system with active compensation for both polarization and phase drifts. The compensation scheme gives an averaged QBER of 0.87% and an averaged duty cycle of 99.6% for a continuous quantum key distribution session of 19 hours over a 20.3km installed telecom fiber [25]. This system seems to be very well developed and also will enter the market soon.

Also in 2005, NEC published a fortnight-long, continuous quantum cryptography final-key generation at an average rate of 13 kBits/s over a 16km-long commercial optical network [26].

In 2005 an improved QKD test system operating at clock rates of up to 2 GHz using a specially adapted commercially-available silicon single-photon counting module was presented. The use of the lossy wavelength of 850 nm for optical fibers reduces the secret bit rates by many orders of magnitude to the kbit/sec level. The use of an enhanced detector had improved the fiber-based quantum key distribution test system performance in terms of transmission distance and quantum bit error rate [27]. In 2006 the system was further improved and a 3.3 Gigahertz clocked quantum key distribution system was presented [28].

In March 2005 the first experimental demonstration of decoy state QKD, over 15 km of Telecom fibers was published. The result showed that, with rather simple modifications to a commercial QKD system, decoy state QKD allowed to achieve much better performance (in terms of substantially higher key generation rate and longer distance) than what was otherwise possible [8]. Further experiments and simulations confirm the advantages of decoy state quantum key distribution [9].

Experiments with Quantum Networks. In 2004 a QKD network between BBN Technologies, Harvard University and Boston University was established under DARPA sponsorship [29, 30]. In 2005 the DARPA quantum network was extended with free-space quantum cryptography by a partnership with QinetiQ [31] and an entangled states source from the Northwestern University was also included.. Moreover, the same institute delivered another quantum communication system that was able to transfer an encrypted messages carried in an OC-3 (155Mb/s) SONET standard between the two nodes.

In April 2004 the integrated project SECOQC started. It is funded by the EU with 11.4 million Euro. The project objective is the development of a global network for secure communication based on quantum cryptography [7]. In contrast to the DARPA network, the founding of this applied research was devoted to many different research groups to generate a broad base for quantum technologies. Never the less the technical goal is to manifest a working quantum network made of at least 6 links to show the typical network structure. First attempts towards standardization are also part of this project.

2.3. Current Practical Limitations

The most obvious practical limitations of current QKD systems are the achievable key rates, transmission distances and system availability. Quantum key distribution is possible over about 100 km standard optical fiber with a key rate of some hundred bits per second. So, today the usage of fiber-based QKD is restricted to metropolitan areas. Free-space QKD systems depend on weather conditions which limits the availability of such systems.

3. QC Application Scenarios in Mobile Networks

Mobile communication networks (see e.g. 3GPP [32]) require a high level of security. To fulfill these requirements end-to-end, mobile network vendors implement state-of-the-art security technologies from the mobile phone trough radio links, core networks, and back-office systems. Siemens Communication's Mobile Networks division has recently reviewed the applicability of quantum cryptography for securing its networks and solutions. This chapter summarizes the key findings.

3.1. Short-term Applications

While algorithmic cryptography and quantum cryptography are often perceived as alternatives, there are also synergies. A good example is the usage of quantum physical random generators in combination with cryptographic algorithms. In contrast to algorithms, such quantum physical random generators provide a perfect solution for generating random numbers required for the generation of cryptographic keys. Due to the commercial availability of such products, that technology is ready to use [33].

3.2. Mid-term Application Scenarios

In the mid-term, we foresee a broad range of application scenarios in mobile networks, where network security will benefit from quantum security technologies, mainly as a replacement for dedicated connections by VPNs enhanced by QKD. Also in mobile networks, the adoption of such innovations will depend on:

- Security requirements.
- Cost considerations for technologies with comparable performance.
- Maturity of quantum computing.

Security requirements for mobile networks always increase as also attackers innovate. Cost for commercial technology is early-on very high, but decreases with increasing adoption and market share. So these two factors will predictably lead to a slowly increasing number of commercially viable applications for quantum cryptography. The advent of quantum computing, while no longer science fiction (see chapter 4.6), is expected to lead to a disruptive increase in the adoption of quantum cryptography, as many other cryptographic technologies will then become ineffective.

Connecting Subscriber Data Bases. Within the core of mobile networks, there are several systems incl. large data base systems with large amounts of subscriber data, e.g. the 2G Home Location Register (HLR) or the Home Subscriber Service (HSS) of 3GPP. The availability of the HSS is critical for the operation of mobile networks. Mirroring of the HSS data is one option to improve its availability. A classical HLR consists of 2 clusters, each holding about 50% of the subscriber records, with typically 60GB of data per HLR. Such a "mated pair" is located in geographically different locations, e.g. within a metropolitan area.

The mated pairs have to be synchronized for any changes. Synchronization should be performed e.g. every 15 minutes. Security requirements regarding confidentiality and integrity are high. This scenario may soon be ripe for the application of QKD for interconnecting subscriber data bases.

For inter-urban distances the data rates fit well with the QKD performance already available, so the commercial feasibility depends mainly on the security requirements and cost/performance ratio in comparison to classical VPN solutions. For inter-urban distances, improvements regarding QKD data rates may be necessary, depending on the security requirements and the resulting frequency of key exchange.

Transport of Accounting/Charging Information. Another mid-term application scenario for QKD-enhanced VPNs is the transport of accounting and charging information from a national mobile communication network to its back-office billing systems. Both the subscriber and the network operator require correct, timely processing of these data, so the security (integrity, proof-of-origin,...) of charging data is critical.

One method for the transport of charging tickets relies on FTP push or pull, with typical data sets comprising around 100 kbit/s. FTP pull is required with a frequency from once per day/week up to once per 10 minutes.

Today, cost/performance ratio of QKD and the distances between national nodes of the mobile network still limit the commercial feasibility. As soon as distances of some hundred kilometers are supported, this becomes a potential application scenario for QKD.

Interconnections between Mobile Communication Networks. Quite often, the partners communicating with their mobile phones are customers in different mobile networks. So there are interconnections necessary between these networks. Typical data here include data rates per individual connection of between 30 and 180 kbit/s, depending on the type of content (voice/data) partly with strict real-time requirements. Per million subscribers some 10.000 inter-network connections have to be supported.

Given the global/nation-wide distances between single networks, the amount of connections which require separate, individual protection, and the real-time requirements for voice, this application scenario poses some extra requirements for QKD.

4. Research Challenges and Recent Advances

4.1. Distance and Rate Limitations of QKD

One of the main goals of the ongoing scientific work of QKD is to increase the distance and rates of single links. Thereby the current limitation seems no longer to be the source of single photons, but the detection of these single photons at a typical communication wavelength of 1310 nm or 1550 nm. The maximal distance of any QKD system (even with perfect sources) is approximately given, when the incoming photons are roughly in the order of the noise of the detector due to dark counts and after-pulsing-effects. The number of incoming photons is limited by the loss of the transmission system and the production rate of the source. Unfortunately the latter value can not longer be increased, because the used detectors are needed to be gated with a maximal rate of 4-10MHz [16]. For practical QKD systems the actual key rate is a factor of 100-1000 smaller, depending on the QKD scheme and the protocol that are used [18, 24, 25, 27].

4.2. Wavelength Conversion Circumvent Detector Problems

The brand new field of wavelength conversion is a research focus in quantum communication [34]. The single photon is converted from 1310 nm or 1550 nm to a wavelength that is possible to detect by some avalanche photo diodes in the visible or near infrared. These devices have a very high detection efficiency of >65% (instead of >10% for 1550nm) and also don't need to be gated. Their low back ground is very preferable for long-distance demonstrations [35]. Even in that first demonstration a secure key rate of 20kHz over a distance of 50 km was presented. Another advantage is the possibility of the transfer to a wavelength that could be used for Ion-traps to act as a quantum memory and to store a single photon. This is a pre-requisite for future quantum repeaters (see 4.5).

4.3. Quantum Networks

As soon as the number of employed stations increases, the optimal architecture with lowest costs changes from single links to quantum networks. Here, only one additional single link is necessary to connect an additional station to the whole network and subsequently to all other users attached so far.

If a sender A and a receiver B would be connected via a network together, then a key is exchanged between the quantum network and sender A and also another one between the quantum network and receiver B. The network has both keys and so the full information of the secrets between A and B. Subsequently, the largest disadvantage of this concept is that the nodes of quantum network need to be trusted.

One strong argument to cut costs in the construction of future quantum architectures is to use existing fiber infrastructure like employed mash and ring networks. Subsequently, the future quantum network structure must be based on very universal quantum architectures. On the other hand, too many requirements would limit the usability, so it is maybe preferable to build up a quantum network starting from scratch. Then also considerations of efficiency of links and different QKD systems can be included.

4.4. Quantum Network towards High Resistance against Attacks

If a sender A and a receiver B of a single link would like to communicate with each other, but the adversary E changes the situation that A and E build up a secret key, but also E and B another one, he acts like a "man in the middle". Quantum cryptography alone is not able to resist this attack. As in classical considerations, only if A and B use a pre-shared secret to authenticate some needed messages in the quantum protocol before a new secret can be established, then the adversary E is unable to perform this "man in the middle attack".

On the other hand, this attack leads to a reduction of this pre-shared secret and the adversary could continue his attack until no secret is left. The stations A and B could not restart the quantum link any longer. This "denial of service attack" is impossible if this link is part of a multi-path quantum network with more than one possible connection between A and B. The stations A and B would be able to exchange a new shared secret between them over the remaining network that has not been tampered by the adversary.

4.5. Quantum Repeater

With the help of amplifiers, the distance of classical communication can be increased by restoring the amplitude of the signals. Contrarily, quantum signals are not able to be amplified due to the same fundamental reasons that ensure security. Nevertheless another method is in principle possible to increase the transmission distances: quantum repeater on the basis of entanglement swapping [36].

If a station A would again like to communicate with a station B, but the distance would be slightly too long for a single link, it is possible to distribute two entangled pairs (a1-b1 and a2-b2) to them in such a way that the photon a1 from the first pair would be received at station A and the photon b2 from the second pair at the station B. Both remaining pair photons (b1 and a2) would need to be saved in quantum memories in a repeater station C in between. The outcome of a joint measurement of b1 and a2 is needed by the stations A and B to form the same secure key out of the measurement of a1 and b2, respectively.

Subsequently it is possible to have more than one repeater in between A and B. In contrast to the quantum network all of them don't need to be trusted, because there is no quantum information transferred into classical bits.

A long-life-quantum memory that was demonstrated in 2005 was an important step towards that direction [37]. Never the less, a quantum repeater can be treated as a small quantum computer with an estimated demonstration horizon between the years 2015 and 2025.

4.6. Quantum Computing

Very recently for the first time even eight ions were entangled with each other [38]. Also the photon/ion interface was published one week later. Both works have been published in "Nature" and are a strong indicator for the progress towards realizing a quantum computer.

Quantum computing is attracting a substantial amount of research funds (e.g. 100 million US$ by Darpa [39]). While there are still many problems unsolved, in summer 2005, a first commercial company, D-Wave, has been founded for the development of a quantum computer [40]. Additional founding will be raised as soon as quantum effects will limit the further development of electronics like processors or memories. Practical quantum computing may still be quite some years away, but it should no longer be considered as science fiction.

4.7. Security of Quantum Cryptography Systems

Most of the nowadays used cryptographic algorithms rely on the fact, that it is computationally hard to find the secret key; such algorithms are rated as computationally secure. But during its evolution cryptography has often found weak spots in cryptographic algorithms or their implementations. Unconditionally secure cryptographic algorithms, which have been proven secure independent of the computing power of an adversary, are an alternative to fulfill very high security requirements.

Assuming that our understanding of quantum physics is correct, and assuming perfect implementation, then quantum key distribution (QKD) has been proven to be unconditionally secure. But QKD alone does not provide a secure communication system. As explained before, two more requirements have to be fulfilled to enable secure communication: authentication to ensure proof of origin and message integrity, and message encryption. There is only one unconditionally secure method for encryption, namely the Vernam cipher or modulo 2 one-time pad, requiring keys as long as the message. For authentication, there are unconditionally secure message authentication codes of Wegman-Carter [41, 42]; requiring a pre-established key. This is also required for all other symmetric-cryptography based authentication methods. Such, it is possible to extend QKD to an unconditionally secure communication system (even secure against attacks by quantum computers), but at a certain price. A discussion of the pros and cons of such a QKD-based system and its alternatives (especially symmetric hierarchical systems) is provided in a paper written by Paterson, Piper, and Schack [43]. In December 2005, a workshop has discussed both classical and quantum cryptography [44].

Quantum computing is expected to challenge most of the cryptographic methods in use today, e.g. to break asymmetric cryptographic algorithms in polynomial time [45, 46]. For their future use all these algorithms have to be proven to be resist against attacks from quantum computers. Just migrating all implementations to better methods is considered a huge task in itself, which will take years. But cryptography will be able to cope with that situation. While for symmetric algorithms a doubling of the key lengths seems to be sufficient for most applications, the situation for asymmetric algorithms is more complex. A candidate for improvements over today's RSA-algorithms is e.g. SFLASH (new version) by Patarin [47]. Quantum public-key systems have been proposed [48], as well as quantum digital signatures, where the quantum state of a string of quantum bits is used as a key [49, 50]. There is an ongoing research project systematically investigating asymmetric cryptography methods able to resist quantum computer's attacks [51].

Therefore, QKD, quantum public-key systems and classical cryptographic methods should not be perceived as alternatives, but as synergistic contributions to the task of ensuring secure communications under increasingly powerful attacks. In most cases, also economical and practical aspects have to be taken into account.

5. Conclusion

Quantum cryptography like most of quantum physics is often considered as an arcane topic, its active research community is rather small and tightly knit. But recent experiments and first commercial products have attracted the attention of various industries. Its current constraints, both regarding technological aspects (e.g. distance and data rate) and the rather high cost of its first commercial products, limit still its practical applicability.

But driven by the ever improving technologies of attackers and by the prospects of quantum computing getting closer to reality, several industries like defense, finance, communications are becoming early adopters. Industry's interest and the high level of both public and private funding are pushing forward the state-of-art of quantum cryptography.

It is expected, that the advent of quantum computing will lead to a transition to unconditionally secure cryptographical algorithms, including the broader adoption of quantum key distribution.

A last word of caution: even algorithms proven to be unconditionally secure may not withstand all attacks, because it cannot be avoided totally that practical implementations of such algorithms can be successfully attacked, e.g. by a so-called social-engineering attack exploiting some human weakness.

Acknowledgements

The authors would like to thank Kurt Eder for his support regarding the application scenarios, Thomas Loruenser and Momtchil Peev for their support regarding the state-of-the-art of QKD, Markus Dichtl and Joern Müller-Quade for their advice regarding some cryptographic aspects of this paper. This work was partly supported by the EC-Integrated Project SECOQC, Project Number: FP6-2002- IST-1 -506813.

References

[1] Bennett, C. H.; Brassard, G.: Quantum Cryptography: Public Key Distribution and Coin Tossing. In: *Proc. Intl. Conference on Computer Systems and Signal Processing,* Bangalore, 1984, 175.

[2] Gisin, N.; Ribordy, G.; Tittel, W.; Zbinden, H.: *Quantum cryptography. Rev. Mod. Phys.,* 2002, 74, 145–195.

[3] Dušek, M; Lütkenhaus, N.; Hendrych, M.: *Quantum Cryptography.* In Progress in Optics, edited by E. Wolf (Elsevier, New York, 2006), 49, 381-454.

[4] Gamow, G.: Mr Tomkins in paperback: Comprising 'Mr Tomkins in Wonderland' and 'Mr Tomkins Explores the Atom'. Cambridge University Press, re-issue edition 1993.

[5] Zeilinger, A.: *Einsteins Schleier.* Goldmann 2005.

[6] Al-Khalili, J.: *Quantum - Moderne Physik zum Staunen.* Spektrum Akademischer Verlag Heidelberg, 2005.

[7] SECOQC - *Development of a Global Network for Secure Communication based on Quantum Cryptography,* EU Sixth Framework Programme. http://www.secoqc.net/

[8] Lo, H.-K.; Ma, X.; Chen, K.: Decoy State Quantum Key Distribution. *Phys. Rev. Lett.*, 2005, 94, 230504.

[9] Zhao, Y.; Qi, B.; Ma, X.; Lo, H.-K., Qian, L.: Simulation and Implementation of Decoy State Quantum Key Distribution over 60km Telecom Fiber. 2006. *Proceedings of IEEE International Symposium on Information Theory* 2006, 2094-2098. http://arxiv.org/pdf/quant-ph/0601168

[10] Kraus, B.; Gisin, N.; Renner, R.: *Lower and Upper Bounds on the Secret-Key Rate for Quantum Key Distribution Protocols Using One-Way Classical Communication.* Phys. Rev. Lett., 2005, 95, 080501.

[11] *Quantum Entanglement.* $http://en.wikipedia.org/wiki/Quantum_{e}ntanglement$

[12] Poppe, A.; Hübel, H.; Lederer, T.; Fedrizzi, A.; Vanner, M.R.; Zeilinger, A.: Detection of Polarization Entanglement after 75km of Fiber transmission. *European Conference on Optical Communication (ECOC)* in Cannes from 24th-28th September 2006.

[13] Grosshans, F.; Assche, G V.; Wenger, J.; Brouri, R.; Cerf, N. J.; Grangier, P.: Quantum Key Distribution Using Gaussian-modulated Coherent States. *Nature*, 2002, 421, 238.

[14] Grosshans F.; Grangier F.: *Reverse reconciliation protocols for quantum cryptography with continuous variables.* http://arxiv.org/abs/quant-ph/0204127

[15] http://www.magiqtech.com/

[16] http://www.idquantique.com/

[17] http://www.smartquantum.com/

[18] Poppe, A.; Fedrizzi, A.; Ursin, R.; Böhm, H. R.; Lörunser, T.; Maurhardt, O.; Peev, M.; Suda, M.; Kurtsiefer, C.; Weinfurter, H.; Jennewein, T.; Zeilinger, A.: Practical Quantum Key Distribution with Polarization Entangled Photons. *Opt. Express,* 2004, 12, 3865–3871. http://www.opticsexpress.org/abstract.cfm?URI=OPEX-12-16-3865

[19] $http://www.quantenkryptographie.at/Presseinformation_{P}RODEQUAC.pdf$

[20] Hübel, H.; Vanner, M. R.; Lederer, T.; Blauensteiner, B.; Lorünser, T.; Poppe, A.; Zeilinger A.: High-fidelity transmission of polarization encoded qubits from an entangled source over 100 km of fiber. *Opt. Express*, 2007, 15, 7853-7862. http://www.opticsinfobase.org/abstract.cfm?URI=oe-15-12-7853

[21] Peng, C.-Z.; Yang, T.; Bao, X.-H.; Zhang, J.; Jin, X.-M.; Feng, F.-Y.; Yang, B.; Yang, J.; Yin, J.; Zhang, Q.; Li, N.; Tian, B. L.; Pan, J.-W.: Experimental Free-Space Distribution of Entangled Photon Pairs Over 13km: Towards Satellite-Based Global Quantum Communication. *Phys. Rev. Lett.,* 2005, 94, 150501.

[22] Resch, K. J.; Lindenthal, M.; Blauensteiner, B.; Böhm, H. R.; Fedrizzi, A.; Kurtsiefer, C.; Poppe, A.; Schmitt-Manderbach, T.; Taraba, M.; Ursin, R.; Walther, P.; Weier, H.; Weinfurter, H.; Zeilinger, A.: Distributing Entanglement and Single Photons through an Intra-city, Free-space Quantum Channel. *Opt. Express,* 2005, 13, 202-209. http://www.opticsexpress.org/abstract.cfm?URI=OPEX-13-1-202

[23] Ursin, R.; Tiefenbacher, F.; Schmitt-Manderbach, T.; Weier, H.; Scheidl, T.; Lindenthal, M.; Blauensteiner, B.; Jennewein, T.; Perdigues, J.; Trojek, P.; Ömer, B.; Fürst, M.; Meyenburg, M.; Rarity, J.; Sodnik, Z.; Barbieri, C.; Weinfurter, H.; Zeilinger, A.: Free-Space distribution of entanglement and single photons over 144km. *Nature Physics*, 2007, 3, 481 - 486. http://arxiv.org/abs/quant-ph/0607182

[24] Gobby, C.; Yuan, Z. L.; Shields, A. J.: Quantum key distribution over 122km standard telecom fiber. *Applied Physics Letters,* 2004, 84, 3762-3764.. http://www.toshiba-europe.com/research/crl/QIG/pdfs/scientific-publications/qkd_apl84_2004.pdf

[25] Yuan, Z. L.; Shields, A. J.: Continuous Operation of a One-way Quantum Key Distribution System over Installed Telecom Fibre. *Opt. Express,* 2005, 660, 13, no. 2. http://www.opticsexpress.org/abstract.cfm?URI=OPEX-13-2-660

[26] http://www.theregister.co.uk/2005/06/02/nec_quantum_crypto/

[27] Gordon, K. J.; Fernandez, V.; Buller, G. S.; Rech, I.; Cova, S. D.; Townsend, P. D.: Quantum Key Distribution System Clocked at 2GHz. *Opt. Express,* 2005, 13, 3015–3020. http://www.opticsexpress.org/abstract.cfm?URI=OPEX-13-8-3015

[28] Gordon, K. J.; Fernandez, V.; Collins, R. J.; Rech, I.; Cova, S. D.; Townsend, P. D.; Buller, G. S.: *3.3 Gigahertz Clocked Quantum Key Distribution System.* 2006. http://arxiv.org/abs/quant-ph/0605053

[29] Elliott, C.: *The DARPA Quantum Network.* $http://quantum.bbn.com/dscgi/ds.py/Get/File-542/The_DARPA_Quantum_Network$

[30] Elliot, C.: Quantum Cryptography. *IEEE Security & Privacy,* July/August 2004, 57–61.

[31] $http://www.bbn.com/News_and_Events/Press_Releases/05_06_01.html$

[32] Third Generation Partnership Project (3GPP). http://www.3gpp.org/

[33] $http://www.idquantique.com/products/quantis.htm$

[34] Tanzilli, S.; Tittel, W.; Halder, M.; Alibart, O.; Baldi, P.; Gisin, N.; Zbinden, H.: A Photonic Quantum Information Interface. *Nature*, 2005, 437, 116–121.

[35] Thew, R. T.; Tanzilli, S.; Krainer, L.; Zeller, S. C.; Rochas, A.; Rech, I.; Cova, S.; Zbinden, H.; Gisin, N.: Low jitter up-conversion detectors for telecom wavelength GHz QKD. *New J. Phys.*,2006, 8, 32-44. http://www.arxiv.org/abs/quant-ph/0512054

[36] Zukowski, M.; Zeilinger, A.; Horne, M. A.; Ekert, A. K.: Event-ready-detectors Bell Experiment via Entanglement Swapping. *Phys. Rev. Lett.*, 1993, 71 4287–4290.

[37] Häffner, H.; Schmidt-Kaler, F.; Hänsel, W.; Roos, C. F.; Körber, T.; Chwalla, M.; Riebe, M.; Benhelm, J.; Rapol, U. D.; Becher, C.; Blatt, R.: Robust Entanglement. *Appl. Phys. B,* 2005, 81, 151–153.

[38] Häffner, H.; Hänsel, W.; Roos, C. F.; Benhelm, J.; Chek-al-kar, D.; Chwalla, M.; Körber, T.; Rapol, U. D.; Riebe, M.; Schmidt, P. O.; Becher, C.; Gühne, O.; Dür, W.; Blatt, R.: Scalable Multiparticle Entanglement of Trapped Ions. *Nature*, 2005, 438, 643–646.

[39] Reuscher, G.; Holtmannspötter, D.: *Quantenoptik - Anwendung nichtklassischen Lichts - Technologieanalyse.* Zukünftige Technologien Consulting der VDI Technologiezentrum GmbH, (p. 96 on DARPA funding for Project QuIST), January 2004. See also http://www.darpa.mil/ipto/programs/quist/

[40] D-Wave - The Quantum Computing Company. http://www.dwavesys.com/

[41] Carter, J. L.; Wegman, M. N.: Universal classes of hash functions. *Journal of Computer and System Sciences*, 1979, 18, 143–154.

[42] Wegman, M. N.; Carter, J. L.: New hash functions and their use in authentication and set equality. *Journal of Computer and system Sciences,* 1981, 22, 265–279.

[43] Paterson, K. G.; Piper, F.; Schack, R.: *Why Quantum Cryptography?* 2004. http://arxiv.org/abs/quant-ph/0406147

[44] *Workshop on Classical and Quantum Information Security, California Institute of Technology,* Dec 15-18, 2005. http://www.cpi.caltech.edu/quantum-security/

[45] Shor, P. W.: *Polynomial-Time Algorithms for Prime Factorization and Discrete Logarithms on a Quantum Computer*, 1995. http://arxiv.org/abs/quant-ph/9508027

[46] Wheatman, V.; McRory, L.: *Quantum Computer: The End of Public-Key Cryptography?* 4 Jan 2002, Gartner Research Note.

[47] Patarin, J.; Goubin, L., Courtois, N. T.: C*+- and HM: variations around two schemes of T. Matsumoto and H. Imai. In: *Proceedings of Asiacrypt '98, 1998, Lecture Notes in Computer Science*, 1514, 35–49, Springer Verlag.

[48] Okamoto, T.; Tanaka, K.; Uchiyama, S.: Quantum Public-Key Cryptosystems. In: *Proc. of CRYPTO* 2000, 2000, *LNCS,* 1880, 147-165, Springer Verlag.

[49] Gottesman, D.; Chuang, I. L.: *Quantum digital signatures*. 2001. http://arxiv.org/abs/quant-ph/0105032

[50] Mueller-Quade, J.: Quantum Pseudosignatures. *J. Mod. Opt.,* 2002, 49 (8), 1269–1276.

[51] Project Public Key Cryptography and Quantum Computing (QPKC). http://www.cdc.informatik.tu-darmstadt.de/research/forschung.html

In: Long-Term and Dynamical Aspects of Information... ISBN 1-60021-912-8
Editors: Andreas U. Schmidt et al, pp. 113-124 © 2007 Nova Science Publishers, Inc.

Chapter 9

Why Digital Signatures Fail - Legal Concepts for Long Term Validity in Austria, Germany and Poland

Dariusz Adamski*
The Research Centre for Legal and Economic Issues of Electronic Communication, University of Wrocław
ul. Uniwersytecka 7/10, 51-136 Wrocław, Poland
***Mirosław Kutyłowski*† *and Anna Lauks*‡**
Institute of Mathematics and Computer Science, Wrocław University of Technology, ul. Wybrzeże Wyspiańskiego 27, 50-370 Wrocław, Poland

Abstract

It is often believed that economic failure of electronic signatures is due to the cost of this technology, high demands on the users, and implementation difficulties. However, the main problem is probably neither technology nor implementation issues, but legal situation of cryptographic technology and discrepancy between technical reality and the model adopted by the legal systems.

We analyze models of validity of electronic documents in European law and compare them with technical reality. We show that there is substantial discrepancy between legal approaches (both in theory and practice) and undeniability of electronic documents in mathematical sense. We propose some basic rules that have to be obeyed in order to guarantee compatibility of legal and technical security of electronic documents.

1. Introduction

The main function of written form of expressing will is to confirm legal actions of the signer. Written form provide durability and undeniability of what is put down on the paper

*E-mail address: dadamski@prawo.uni.wroc.pl
†E-mail address: Miroslaw.Kutylowski@pwr.wroc.pl
‡E-mail address: Anna.Lauks@pwr.wroc.pl

and authorized with a handwritten signature. Therefore it establishes an immanent relation between a signer and the content of his will.

Recently, we observe a growing tendency that parties to legal actions can often decide otherwise on how the formal relation between a person pursuing a legal action and his expression of will should be couched. Moreover, quite often laws leaves the decision on how (and if) to evidence a legal action in case a dispute between the parties occur. The legal environment for digital equivalence to handwritten signatures should recognize this desired flexibility and evolution of practice of legal actions. Unfortunately, the current legal models do not seem to appropriately respond to this need. Moreover, general liberalization of legal systems does not affect electronic form, for which a quite restrictive approach is a general rule.

We analyze the legal framework of three European countries - Austria[2, 3], Germany[4, 5] and Poland[6]. Comparing the legal approaches with the technical reality and mathematical background of security of digital signatures we show problems of the legal models, which are currently in use, and their consequences for the long-term validity of digital documents and their signatures. We conclude by an overview of some basic rules that can be used in order to guarantee compatibility of legal and technical security of electronic documents, flexibility of the systems and technological neutrality.

1.1. Organization of the Paper

In Section 2. we describe mathematical issues concerning security and validity of digital signatures, as well as some technical background and available features of digital signatures. Section 3. is devoted to the model of validity of electronic signatures adopted in the legal framework of the European countries concerned. In Section 4. we show the consequences of enforced models for archiving of digital data. In Section 5. we present some rules that would guarantee compatibility of legal rules and technical security of electronically signed documents.

2. Electronic Signatures from the Technical Point of View

2.1. Forms of Electronic Signatures

According to the EU Directive [1] electronic signature can be either attached to the signed data or logically connected to it. This formulation corresponds to the method of deriving a signature of data M by computing $H(M)$ for a hash function M, and then computing signature $s = \mathrm{sig}(H(M),K)$, where K is a private key. Then obviously s can be either attached to M or stored/sent separately from M.

This formulation hardly admits even such classical methods as digital signature with message recovery. Recall that a signature of M is created in this way equals $E_K(R(M))$, where K is a private key, R is a redundancy function, and E is an encryption function. In this case, the message M may not exist outside the signature. So the signed data M has virtual nature, it may not exist at a certain point and later be recovered from its signature. Even if this has quite obvious technological context, it may happen that legal regulations demand storing data in the original form together with appropriate data securing its integrity.

A quite different situation occurs for robust digital watermarks that serve the purpose of protection of data such that digital pictures. In this case the signature changes the original data in a way that it does not change quality of the digital picture from a human point of view, and the signature remains hidden from the observer. Moreover, it should be hard to remove a watermark from the picture. So a signature of this kind is neither attached nor logically connected to the signed data - it changes it and remains hidden, unless revealed with special procedures.

2.2. Signer

From technological point of view it is not the case that a signature can be attributed to only one person (as it is the case for a handwritten signature).

First, *proxy signature* technique [14, 24, 13] makes it possible to delegate the right to generate a signature to a third party. However, such a signature, when created, often cannot be distinguished as a proxy one. So, one cannot be sure that a given signature has been really generated by the owner of the private key - may be he only delegated the right to sign some data to a third party without revealing his private key.

A *group signature* [10, 9] hides from the holder of a signed document who was actually the signer. Nevertheless, it can be checked that this was a person from a certain group. For signature creation only one member of such a group is enough. However, in case of group signatures it is possible to uncover the signer. Usually the appropriate procedures have to be executed by a group *administrator* (for this purpose the group administrator holds some secret keys). In some schemes [15] the function of administrator of the group does not exist and in this case the members of the group can trace the actual signer.

A *ring signature* [20, 11] is a construction similar to group signatures — a signed document indicates that somebody from a declared group (so called *ring*) is the signer. While it is possible to check that this is true, it is impossible to point to the person from the ring who actually signed the document. This is not due to lack of an administrator in the sense of group signatures. Simply, it is impossible from information-theoretic point of view, even if the signature schemes are broken or the keys are revealed. Since creating a ring does not require any cooperation with the people added to the ring, this is an attractive tool for prosecutions with anonymous witnesses.

Legal procedures require often that a document becomes valid if at least two parties sign it. Procedure of signing electronically in a classical way has certain disadvantages — it is impossible to sign at the same time. In some sense *mediated signatures* [8] solve this problem — a single signature for two (or more) parties is created. Generation of such a signature requires usage of private keys of all parties, and intermediate result has no proof value for any external party. Interestingly, RSA-based mediated signatures do no require signing in any particular order — the result is valid anyway.

2.3. Verification Procedures

Legal models of validity of digital signatures almost always assume that everyone who knows the algorithm, its parameters and a suitable public key is able to check the signature validity at any time. However, this is only one line of possible design. *Undeniable signatures* require cooperation with the signer — within a special procedure the alleged signer either proves that she is the signer of a document concerned, or just the opposite. Moreover, a transcript of such a protocol execution is mathematically worthless to a third party. This feature makes undeniable signatures attractive for certain applications, where the signer wishes to keep track who can check validity of a signed document.

2.4. Security of Signatures vs. Security of Private Keys

We have to be aware that from the technological point of view the validity of a signature is not a proof that a certain person has signed the document. The verification procedure tells us only if the signature was created by **a person having the appropriate private key**. The process of assigning and linking the private key with a certain person takes place outside the signing algorithm. Usually the secret key is set up on a special cryptographic card. Nevertheless the private key can always be stolen. The mathematical model of a signature does not have a cure for that situation. For every digital signature scheme it is assumed that a private key has to be kept in secret.

2.5. Randomness vs. Security of Signatures

Keeping secret keys safely is a necessary but not sufficient condition for trustworthiness of electronic signatures. One of the weakest points in the system are random parameters used often by the signature schemes (a major exception is RSA scheme, but even then random parameters are used during key generation and the same problems may arise).

First of the problems is that poor randomness may lead to implementations that can be broken regardless of security of the scheme in the general case [17]. Apart from weak generators of (pseudo)random parameters, there is a whole collection of cryptographic methods that use random parameters to leak information through either subliminal or kleptographic channels [22, 23]. Due to strong cryptography used, detecting such a channel is practically infeasible. In particular, an allegedly secure signature creation device may leak secret keys within signatures created and there is no practical way to check it without having access to internal parts of the device. However, a device should physically prohibit such an access, since it may lead to stealing secret keys as well.

2.6. Proof Value of Digital Signatures

Security and reliability of the signature is usually based on a computational assumption that there is no efficient algorithm that allows to solve a particular problem (for example the ElGamal signature relies on the discrete logarithm problem, the RSA signature relies on the factorization problem).

It should be stressed that the objective proof value of a digital signature decreases in time. It depends on technology available and declines as technology advances. Reasons of that conjuncture are technological development and progress in cryptanalysis. Experience shows that predictions related to the rate of decline are of little value, especially in the long term case. For example in the year 1983 Ronald L. Rivest et al. made a statement [19] that RSA *an 80-digit n provides moderate security against an attack using current technology* and *using 200 digits provides a margin of safety against future developments*. 200 digits n

is less that 670 bit long. The key length is very important from the security point of view and it is known that at present 2048 bit n's are recommended. Even this recommendation may fail to be appropriate, our past experience shows that development of cryptanalysis methods is highly unpredictable.

2.7. Fail-stop Signatures

In the literature one can find signing schemes, so called *fail-stop signatures* [18, 12], that could be helpful to overcome problems mentioned in Section 2.6.. Namely, a fail-stop signature scheme allows a signer to protect herself against solving the underlying hard computational problem by a third party. In case of a forgery (the signature passes the verification procedure, but was not created by the legitimate signer), the fail-stop signature scheme allows the original signer to prove with an overwhelming probability that an underlying computational assumption has been broken and so the forgery has become possible.

3. Traps of Legal Models for Electronic Signatures

Legal rules in many countries require to use an advanced digital signature verified with a qualified certificate whenever the written form requirement has to be fulfilled. In face of problems with certificates that are mentioned in Section 3.2., service provided by a trusted third party (Certification Authority) confirming identity of a signer and his public key becomes quite costly and regardless of the investments not fully trustful. Moreover, such measure seems to be unnecessary in some cases, for example when the signer is a public person for which the public key is published in a different way (like printing the public key in a newspaper).

3.1. Faulty Lists of Secure Algorithms

Many European countries (e.g. Germany, Austria and Poland) try to guarantee security of electronic signatures by establishing within a legal framework suitable lists of algorithms and their parameters that should be used in the procedures of signatures creation and verification. These list have a limited time horizon, typically they are valid for a few years only. Even with this restriction, it leads inevitably to situations that a signature which has lost its proof value in a mathematical and technological sense still is regarded as fully trustful according to the legal system concerned.

The second problem is that closed lists of secure method that can be used for legal purposes limit or even eliminate possibility of using some secure technologies (i.e. fail-stop signatures described in Section 2.7.). As a major step backwards, an open list of secure algorithms has been replaced by a closed one in Austria.

In Germany, the problem of evaluating relevance of algorithms was solved differently. According to Annex 1 to the Ordinance on Electronic Signatures[5], publishing an overview of suitable algorithms and related parameters is a task of an appropriate authority. But even shifting the responsibility to a specialized institution did not protect Germany from stating in March 2005 that SHA-1 is a secure and suitable hash function (up to 2010) despite the Chinese attacks on SHA-1 which became known at this time [21]. In Poland, according to the Digital Signature Act[6], composing a list of secure algorithms is a duty of the government and the list becomes a part of the legal system. Regardless of many basic errors contained in [7], no corrections have been implemented so far.

Practical failure of the concept that it is responsibility of the state to determine secure authentication techniques implies necessity to design a new legal framework for this problem. One plausible concept is to apply the rules that are compatible with the rules concerning handwritten signatures. According to them, it is responsibility of the person entrusting a digital document to check whether authentication techniques are trustworthy. Such a person bears the legal risk and therefore should be careful enough. This provides room for free market, evaluation by independent organizations and bodies publishing "black" and "white lists" of available signature methods.

3.2. Validity of Signatures vs. Validity of Certificates

According to the legal framework of the countries concerned validity of secure electronic signatures depends on the procedure of signature verification and validity of the qualified certificate of the signer. There are two main approaches:

1. The signature is valid provided that it can be verified with a valid qualified certificate of the signer (Austria, Poland).

2. The signature is valid provided that in the moment of signature creation a qualified certificate of the signer was valid (Germany).

If a digital document has to preserve its validity for a period longer than the period of the certificate's validity, we get problems.

According to the first model a signer should get a new certificate and re-sign the document, before a certificate becomes invalid. Moreover, in Poland, it is impossible to get a new certificate for the previous public key with the same identifier. Since the certificate used for signature verification has to be explicitly indicated in the signature, there is no possibility to re-fresh the signature of an old document. Apart from this legal trap, re-signing is a costly, time consuming and unnecessary procedure, unfounded from the mathematical point of view. Currently, in Poland the problem is even more acute, each document has to be signed separately, so this excludes possibility of any automation of the re-signing procedure (which could be useful for archiving purposes).

The next serious disadvantage of the first model are potential abuses that are possible through certificate revocation. Indeed, an electronically signed document looses its attribute of a written form when a certificate looses its validity. So the simplest method of denial of signatures is revocation of the certificate. This blocks for instance usage of digital signatures in many court procedures.

The second model seems to be more reasonable. Signer does not have to re-sign each document when the certificate used becomes invalid. The necessity of re-signing (according to the German Ordinance on Electronic Signatures[5]) appears only if *data with a qualified electronic signature are required in signed form for a period longer than that for which the algorithms and related parameters used to create and verify them are considered to be suitable.* The dark side of this approach is that one has to check that a certain qualified certificate was valid at the time when the signature was created. In general, it is not straightforward to establish the time of signature creation; the signature alone cannot prove it with cryptographic means.

3.3. Limiting Usage of Advanced Technologies

One of the major drawbacks of legal rules of what is "secure" is blocking technological advance and use of the most trustful technologies and schemes. For instance, the signatures described in Section 2.2. and 2.7. do not have legal consequences regarding written form, so their practical usage has been hardly limited (or even eliminated).

One of the major mistakes of legal rules of electronic signatures in many countries is that an electronic signature has to be linked to a natural person (Germany, Austria and Poland follow this approach, while Czech Republic abandoned this approach). In this way the most natural application field of electronic signatures has been excluded - namely a fully automatic workflow of legal documents issued by legal persons.

The legal regulations miss the point that signature schemes like proxy, group, mediated, fail-stop signature schemes allow not only to solve some security problems from traditional workflow of paper document, but also offer features that are impossible for the paper documents. Unblocking usage of these schemes would contribute a lot to effectiveness of electronic document processing systems.

4. Consequences for Archiving Digital Data

Arguments mentioned in the previous section show that the German model of validity is better suited for archiving digital documents. It is not necessary to re-sign documents in order to confirm their existence at a certain time. A standard technique used for this purpose are Merkle trees based on secure hash functions [16]. According to this technique we put signed documents into leaves of a Merkle's tree. Then we compute value of a node in the tree as a value of hash of the values of its sons. Proof of existence at a certain time is given by time-stamping the value in the root of a Merkle's tree. Such a solution ensures that any document cannot be changed, deleted or added without reconstructing the whole archive.

The problem that we can not escape in case of long term validity is that the hash functions can be broken. However, we have to be aware that finding a collision gives only a possibility of changing the documents in the leaves of a Merkle's tree (adding or deleting files by a third party from the archive is still impossible as long as inversions of hash functions remain unfeasible). On the other hand, in the leaves of a Merkle's tree we store signed documents, so in order to manipulate an archive an adversary has not to find merely a collision, but a collision with a signed document. This is much harder from computational point of view, since secure signing is a one-way operation.

Archiving digital documents is also influenced by the problem of availability of software necessary to interpret a data format in which documents stored in the archive were created. The problem may appear, if the format used for an electronic document is not open and the support for the format is offered by a single company. A proper solution is to use formats that are either fully defined in widely available documentation or are in some sense self-contained. Additionally we need a right to use the format freely in order to avoid legal problems.

5. Proposed Validity Approach

5.1. Default Rules - Place of Origin

One of the basic rule regarding function of hand written signatures in continental European law is that signing act has certain functions defined within the legal framework of a country. Regulations on electronic signatures simply add additional procedures that have the same effect. Paper documents are usually retained within a country, where a document has been created, and so interpretation of document's legal functions are the same as regarded while creating the document. Electronic data are transfered with much ease, so inevitably this default assumption is no longer true.

A simple rule that can be used to resolve the problem of incompatibility of legal functions of electronic documents in different countries should be the following:

Rule 1: *Legal form of an electronic document is defined by the place where a document has been created. This default rule can be overrun by an explicit declaration in the document stating that the rules of a different place of origin apply.*

5.2. Self-contained Electronic Data

It is often declared that the main function of digital certificates is to determine identity of the owner of a public key. However, in practice it turns out that other functions might be equally important. The first example are limitations on the value of transactions which are signed with the private key corresponding to the public key contained in the certificate. Other examples are given by attribute certificates that state in more detail the rights of the holder of appropriate private keys.

In these cases the role of a certificate is to provide more details on legal value of an electronic signature. In some countries (for example Poland), it is even required by law to incorporate a certificate into a digitally signed document. The aim of such an approach is to provide a framework, in which a document itself determines its legal value. (This goal is hard to be achieved fully, since finally validity of the certificate has to be checked online.)

In order to enable unambiguous legal interpretation of a digital document in automatic or semi-automatic way the following rule has to be obeyed:

Rule 2: *An electronic document should contain a complete list of*

- *all legal conditions concerning its validity, and/or*
- *the links to other documents concerning its validity.*

In the second case, the links should be secured digitally so that they point to a unique document that cannot be altered.

In particular, the link mentioned in Rule 2 can be secured by providing a hash value or an advanced electronic signature for this document.

Rule 3: *The conditions defined according to Rule 2 must define the procedure that enable a holder of the electronic document to evaluate the legal consequences of the document.*

The goal of Rule 3 is to enable algorithmization of legal interpretation of digital documents. Once such procedures are defined then one can assign one of the following values to the document for a certain role: *valid* (if it passes the test), *invalid* (if the procedure rejects the document), *unknown* (if it is not feasible to perform the procedure). It should be possible to refine these test result, if needed.

One of the consequences of the above approach is that a document with an electronic signature fulfills the requirements for written form only if it was explicitly stated. On the other hand, according to European Directive on Electronic Signatures [1], any electronic signature created with a secret key for which there is a corresponding qualified certificate is legally binding as an handwritten signature, even if the signer has different intentions. Even if it is required that a secure signature-creation device should warn the signer about legal consequences, there is no way to check whether this has really occurred.

5.3. Validity Models

In order to facilitate interpretation of legal context of electronic data standards regarding validity models should be determined. The standards should encompass:

- the rules following from the legal framework of the place of origin,
- additional conditions and limitations imposed by the originator of an electronic document.

In particular, standards concerning national legal context of electronic data should be defined. This would help to resolve the problems such as for the classical case of documents with electronic signatures circulating between Austria and Germany.

5.4. Technical and Mathematical Proof Strength

Proof provided by an electronic data can be regarded from different points of view: the legal one and the technical and mathematical one. The general rule should be that:

Rule 4: *A proof provided by an electronic data should be considered in a legal sense as long as the proof is sound in mathematical and technical sense.*

Even if Rule 4 sounds quite obvious, it has been often violated, as mentioned in Section 3.. This general rule has the following consequences:

1. The mathematical and technical evidence is superior to formal rules imposed by authorities (such as validity periods of algorithms), evaluation of algorithms and technologies by supervisory bodies becomes a guideline for the users, independent evaluation and scientific arguments become equally important.
2. Strict validity period of digital certificates becomes an option and not a compulsory measure.

3. The verification procedure is performed according to the objective situation at the time of verification. It may concern all data existing at the time of verification, but also flaws and attacks discovered after creation of electronic data concerned.

5.5. Time and Keys in Electronic Systems and Their Legal Counterparts

As demonstrated in Section 3., simplifications that are present in legal systems may lead to situations, where trust in electronic documents will be endangered. In consequence, electronic documents will be distrusted even if they are legally binding. In order to avoid such situations, the following rules should be implemented:

Rule 5:

- *An electronic signature should be regarded as a proof that a person holding the private key used for creating the signature has signed the document concerned. It should not be the basic principle that the signature was created by a person pointed to by some qualified certificate.*
- *Time in electronic systems should be regarded as in the model of Lamport's logical clocks; traditional notion of time should be confined to local systems; no synchronization should be assumed except that it is explicitly established by some technical means.*

References

[1] Directive 1999/93/EC of the European Parliament and of the Council of 13 December 1999 on a Community framework for electronic signatures. OJ No L 13 p. 12 19/1/2000. (http://europa.eu.int/eur-lex/pri/en/oj/dat/2000/l$_0$13/$l_0$1320000119en00120020.pdf)

[2] Austrian Digital Signature Act: Bundesgesetz über elektronische Signaturen (Signaturgesetz - SigG), BGBI. I Nr. 1999/190. (http://www.signatur.rtr.at/en/legal/sigg.html)

[3] Austrian Signature Ordinance: Verordnung des Bundeskanzlers über elektronische Signaturen (Signaturverordnung - SigV), BGBI. II Nr. 30/2000. (http://www.signatur.rtr.at/en/legal/sigv.html)

[4] German Digital Signature Act: Gesetz über Rahmenbedingungen für elektronische Signaturen und zur Änderung weiterer Vorschriften, Bundesgesetzblatt Teil I 22: 876-84. (http://bundesrecht.juris.de/sigg$_2$001/*index.html*)

[5] Ordinance on Electronic Signatures: Verordnung zur elektronischen Signatur, Bundesgesetzblatt Teil I 59: 3074-84. (http://bundesrecht.juris.de/sigv$_2$001/*index.html*)

[6] Polish Digital Signature Act: The Act of 18 September, 2001 on Electronic Signature. (http://www.mgip.gov.pl/NR/rdonlyres/0BAA71A7-B442-4D61-99C4-51AF277FA9F5/18224/act$_o$n$_e$*Signature.pdf*)

[7] Rozporządzenie Rady Ministrów z dnia 7 sierpnia 2002 r. w sprawie określenia warunków technicznych o organizacyjnych dla kwalifikowanych podmiotów świadczących usługi certyfikacyjne, polityk certyfikacji dla kwalifikowanych certyfikatów wydawanych przez te podmioty oraz warunków technicznych dla bezpiecznych urządzeń służących do składania i weryfikacji podpisu elektronicznego. (Polish Ordiance on Electronic Signatures) (http://www.mgip.gov.pl/NR/rdonlyres/BEF6E08C-F7CB-496A-9B9D-C8AE365D5DD2/0/6020_elektro.htm)

[8] Boneh, D.; Ding, X.; Tsudik, G.; Wong, C. M.: A method for fast revocation of public key certificates and security capabilities. *10th USENIX Security Symposium*, Washington, D.C., 2001.

[9] Camenisch, J.; Michels, M.: A Group Signature Scheme with Improved Efficiency. *ASIACRYPT* 1998, *LNCS* **1514**, pp. 160-174, Springer-Verlag,1998.

[10] Chaum, D.; van Heyst, E.: Group Signatures. *EUROCRYPT* 1991, *LNCS* **547**, pp. 257-265, Springer-Verlag, 1991.

[11] Herranz, J.; Sáez, G.: New Identity-Based Ring Signature Schemes. *ICICS* 2004, *LNCS* **3269**, pp. 27-39, Springer-Verlag, 2004.

[12] van Heyst, E.; Pedersen, T. P.: How to Make Efficient Fail-stop Signatures. *EUROCRYPT* 1992, *LNCS* **658**, pp. 366-377, Springer-Verlag, 1992.

[13] Hwang, M.-S.; Lu, E. J.-L.; Lin, I.-C.: A Practical (t, n) Threshold Proxy Signature Scheme Based on the RSA Cryptosystem. *IEEE Transactions of Knowledge Data Engineering* **15**(6), pp. 1552-1560, 2003.

[14] Mambo, M.; Usuda, K.; Okamoto, E.: Proxy Signatures for Delegating Signing Operation. *ACM Conference on Computer and Communications Security* 1996, pp. 48-57, 1996.

[15] Manulis, M.; Sadeghi, A.-R.; Schwenk, J.: Linkable Democratic Group Signatures. *ISPEC* 2006, *LNCS* **3903**, pp. 187-201, 2006.

[16] Merkle, R. Ch.: *Secrecy, authentication, and public key systems.* UMI Research Press, Ann Arbor, Michigan, 1979.

[17] Nguyen, P. Q.; Stern, J.: Lattice Reduction in Cryptology: An Update. Algorithmic Number Theory Symposium 2000, *LNCS* **1838**, pp. 85-112, Springer-Verlag, 2000.

[18] Pedersen, T. P.; Pfitzmann, B.: Fail-Stop Signatures. *SIAM J. Comput.* **26**(2), pp. 291-330, 1997.

[19] Rivest, R. L.; Shamir, A.; Adleman, L. M.: A Method for Obtaining Digital Signatures and Public-Key Cryptosystems (Reprint). Commun. *ACM* **26**(1), pp. 96-99, 1983.

[20] Rivest, R. L.; Shamir, A.; Tauman, Y.: How to Leak a Secret. *ASIACRYPT* 2001, *LNCS* **2248**, pp. 552-565, Springer-Verlag, 2001.

[21] Wang, X.; Yin, Y. L.; Yu, H.: Finding Collisions in the Full SHA-1. CRYPTO 2005, *LNCS* **3621**, pp. 17-36, Springer-Verlag 2005.

[22] Young, A.; Yung, M.: Kleptography: Using Cryptography Against Cryptography. *EUROCRYPT* 1997, *LNCS* **1233**, pp. 62-74, Springer-Verlag, 1997.

[23] Young, A.; Yung, M.: Malicious Cryptography: Kleptographic Aspects. CT-RSA 2005, *LLNCS* **3376**, pp. 7-18, Springer-Verlag, 2005.

[24] Zhang, K.: Threshold proxy signature schemes. Information Security Workshop 1997, *LNCS* **1396**, pp. 282-290, Springer-Verlag, 1997.

In: Long-Term and Dynamical Aspects of Information... ISBN 1-60021-912-8
Editors: Andreas U. Schmidt et al, pp. 125-139 © 2007 Nova Science Publishers, Inc.

Chapter 10

LONG-TERM SECURITY FOR SIGNED DOCUMENTS: SERVICES, PROTOCOLS, AND DATA STRUCTURES

***Thomas Kunz*[*] *Susanne Okunick*[†] *and Ursula Viebeg*[‡]**
Fraunhofer Gesellschaft (FhG),
Institute for Secure Information Technology (SIT),
Rheinstr. 75, 64295 Darmstadt, Germany

Abstract

Long-term retention of electronically signed documents brings with certain problems: Signed documents may loose their probative value over the years and changing of data formats breaks the original signature. In this paper, we analyze the state of the art to resolve these problems and describe requirements, existing and partially field-tested concepts, data structures and specifications of service protocols. Solutions for the aging problem are already in an advanced state, whereas in the area of secure transformation merely first requirements and solution concepts may be presented. Since many organizations, and, in particular, SMEs and individuals are confronted by the challenge how to technically and organizationally implement existing concepts we describe services caring about the long-term and law-abiding preservation of electronic documents. In this way users of a web service are not forced to deal with the complex process of securing their documents over long periods of time.

1. Introduction

Today, most of the business relevant documents, like invoices, are created, exchanged, and stored in electronic form. The legislation on European level has provided the legal framework for a barrier-free electronic communication and electronic commerce: In 1999 the EU Directive 1999/93/EC [10] was enacted, providing legal requirements for a common introduction of electronic signatures in Europe. The German Signature Law (in German: Signaturgesetz or SigG) transposes this Directive into German legislation. Thereby for certain types of signature, e. g. the so called "qualified" signature, the legal recognition is

[*]E-mail address: thomas.Kunz@sit.fraunhofer.de
[†]E-mail address: Susanne.Okunick@sit.fraunhofer.de
[‡]E-mail address: Ursula.Viebeg@sit.fraunhofer.de

enabled in a way that the legal status of electronic signatures is equivalent to that of hand-written signatures in paper documents. For electronically issued and tax relevant invoices the German Value Added Tax Act (in German: Umsatzsteuergesetz) requires qualified signatures according to the German Signature Law. The legal basis for the usage of electronic signatures is provided also for the public sector within national and community administration, for public health, banking, and the justice system.

Record retention bases on legal requirements for both electronic and paper based documents in the same way. The change to electronic processes increases the need for the secure storage of electronic documents. Long retention periods, e. g. more than 30 years, cause problems especially for electronically signed documents. Two problems regarding the preservation of the probative value of signed documents have to be resolved:

1. Electronic signatures are aging. The increasing computing power, the possibility of networking, and progress of cryptography contribute to the "weakening" of electronic signatures, i. e. electronically signed documents may loose their probative value over the years. This process may already be significant after six years.

2. Electronic signatures break when changing the document format. The technological development, harmonization attempts, and also new juridical guidelines cause changes of user data and signature formats over the years. With electronically signed documents format changes are problematical, since changing the format breaks the original signature. A similar problem arises during the digitalization of paper documents. If, for instance, a document signed by hand is digitized, the signature looses its validity. The legal authenticity of the transformed document is at least doubtful.

In this paper, we first give a state of the art overview of solution concepts for these basic problems. Some concepts concerning the weakening process of signature algorithms exist and there are already standardization efforts. First concepts dealing with secure format changing are developed as well.

Currently many organizations are confronted with the challenges how to lawcompliantly preserve electronic documents for a long period of time and how to technically and organizationally implement the existing concepts. In particular, SMEs (small and medium-sized enterprises) and individuals are often overwhelmed.

Appropriate services for archiving and transformation of documents could produce relief and are of particular interest. To establish such services, consistent service descriptions, standardized interfaces, and interoperable data structures are essential.

Therefore we design some selected services with their architecture and basic tasks, to realize parts of above named concepts, e. g. an archiving service caring about signature renewals, a certification service (in German: Beglaubigungsdienst) or a transformation service for conversion of data formats with subsequent notarization. We identify missing solution concepts and gaps in standardization.

Our goal is to increase the applicability and dissemination in this field of activity. Several archiving problems, such as the long-term readability of storage media and the protection of the documents against deletion or manipulation during the archive period, are not addressed in this paper.

The paper is organized as follows. Section 2 attends the long-term conservation of the probative value of signed documents. In section 3 we deal with the legally secure transformation of signed documents. In section 4 we compare the essential concepts and draw some conclusions. Finally we give a summary.

2. Long-Term Conservation of Signed Documents

In this section, we address the problem of signature aging. The following analyzes the state of the art and provides an introduction to existing concepts and standardization efforts in this field of development. We describe different services realizing these concepts and briefly propose field-tested data structures as a subject for standardization.

2.1. State of the Art

All existing concepts are based on the timely renewal of signatures by means of time stamps. Timely means to renew before the currently used signature algorithms become insecure. A time stamp is issued by a time stamp service, which receives a hash value of data, adds the current time, and signs this generated construct by its own signature based on stronger algorithms or parameters. The time stamp itself is a signed date and thus also subject to an aging process. For this reason it is integrated in the archiving process and continually renewed. In this way chains of time stamps are generated.

In [15] the European Telecommunications Standards Institute (ETSI) specifies signature formats based on CMS and XML signatures and also takes into account longterm conservation of provability of signed documents. By this concept an electronic signature is first completed with verification information (e. g. certificates and certificate revocation lists) needed to verify the signature. The collected data is encapsulated in one object and secured by a time stamp. As we will see later this signature oriented method is only suitable for small numbers of documents.

The IETF (Internet Engineering Task Force) working group LTANS (Long-Term Archive and Notary Services) deals with the same topic and has already defined requirements, data structures and protocols for the secure usage of archive services. The working group is initiated as a result of the project ArchiSig, funded from 2001 to 2003 by the German Federal Ministry of Economics and Technology [4].

For long-term archive services (LTA) which are responsible for the preservation of data over long periods of time the "Long-Term Archive Service Requirements" have been defined [6]. They can be roughly divided into requirements for the access protocol to communicate with the LTA and functionality requirements. A protocol has to support the basic operations: Submit data objects for archiving, retrieve and delete archived data objects, specify, extend or shorten archiving periods, specify metadata associated with a data object, and specify an archive policy under which the submitted data should be handled. An archive service must provide evidence records containing sufficient information for a non-repudiation proof that a data object existed at a certain time. Hereby the evidence record must allow detection of any modifications to it and the appropriate data objects. The service has to operate on the basis of an archive service policy consisting of rules for operational aspects and authorization concepts. It also must be possible to transfer data objects and its evidence record from one service to another without losing evidence value. Furthermore data confidentiality must be ensured between the submitter and the LTA.

The structure of evidence data for an archived data object is closely coupled with the underlying concept [7]. This concept is in contrast to ETSI document oriented and independent of signature formats. First the electronic document as a whole with all its signatures is hashed and time stamped. To optimize the process lots of signed or unsigned documents are merged in a hash tree, first described by Merkle [8], whose root hash value represents all documents. A time stamp is requested only for the root hash of the tree. This procedure is more cost effective and mass-capable. The hash algorithm used for building the hash tree may also become weak, which requires the more complex hash tree renewal by using a current suitable hash algorithm.

The ERS (Evidence Record Syntax) proposed by the LTANS group [1] describes the data structure of an evidence record. It contains all time stamp chains and other verification data necessary for the complete proof of existence of an archived data object or a data object group at a certain time. The ERS takes into account the archive requirements and meets the requirements, in particular, for data structures. It handles huge numbers of data objects and data of any format. The work on ERS is well advanced by now and first supporting software solutions are in the market.

Based on the archive requirements the LTAP (Long-term Archive Protocol) [9] describes basic functions of a protocol for communication between clients and a LTA. These are functions for transferring data to the archive service, accessing to and deletion of archived data as well as verifying the integrity and authenticity of the archived data. The LTAP draft is in an early development stage until now, so it does not define sufficiently concrete data structures for practical use.

Both the ETSI and the LTANS concepts of signature renewal are accepted by ISIS-MTT [16].

2.2. Long-Term Archive Services

As we have seen, requirements for long-term archive services as well as concepts for preserving the probative value of electronic signatures already exist. We have therefore accomplished the prerequisites for the definition of LTAs. Basically we distinguish two kinds of services.

The first type of service is intended for larger organizations or companies. They already run some kind of document management system (DMS) to file their documents. They only need to add an additional service for ensuring the preservation of the probative value of their signed documents. This kind of service is characterized by the integration in an environment with an already existing document repository. We assume, for pure document management users want to use their familiar and already existing DMS clients. The process of long-term securing the documents can be done fully automatically in the background. An interaction with the user is not necessary. If a new document is stored in the document management system, the DMS informs the LTA on this fact. The long-term archive service must have access to this document. Accessing the document means in this context that the service only fetches the document in order to compute the hash value of the document

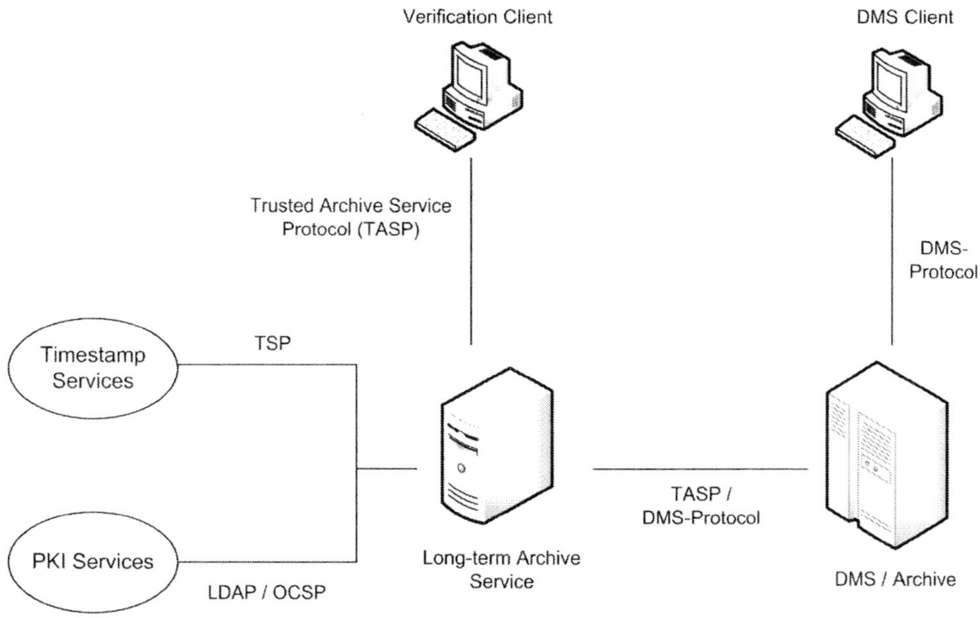

Figure 1. Long-term archive service in combination with a DMS.

and to obtain a time stamp. The document resides unchanged in the DMS. The LTA only holds the corresponding evidence record. Additionally the long-term archive service must offer methods to access the evidence records. Since accessing the evidence records is a rather infrequent task, we propagate a separate client for requesting and verifying evidence records. Thus, the whole process of securing the documents can completely be hidden from every day work. Figure 1 shows a possible architecture with a separate verification client and a DMS client.

The second type of long-term archive service is rather intended for SMEs and individuals who are often overwhelmed by the challenge of preserving the probative value of documents. The service can be imagined as a web service based on technologies like SOAP [12] and WSDL [13]. Alternatively the service can be accessed via a web browser interface. Privacy and confidentiality play a major role than in the other scenario. In general, users do not want to entrust their documents to the service. Hence the service is only used for performing the signature renewal and managing the evidence records i. e. it has no access to the user's documents. Instead the user only provides hash values of the documents to the service. In contrast, as pointed out in [11], the submission of encrypted documents is problematic since the evidence that encrypted data existed at a certain point of time is not necessarily an evidence for the existence of the appropriate unencrypted data. Just as the first service, also this service must offer the possibility to access the evidence records. Unlike the first type of service, here the user is more actively involved in the process. The user not only directly determines which documents have to be secured. It is supposable that the user can also make individual adjustments concerning supported cryptographic algorithms, used timestamp services and so on. That means, every user can force its own policy under which the service should act. Furthermore, the client has to care about the storage of the user's documents and must be able to hash documents. The service will inform the user if the used hash algorithm becomes insecure. In this case the user must re-hash his documents with the currently valid hash algorithm.

Both types of services require a PKI infrastructure (LDAP, OCSP etc.) as they collect certificates and revocation information in order to verify time stamps and, at least in the first scenario, document signatures. The services store all collected PKI artifacts to enable a later verification, even if the PKI services are no longer operating or at least do no longer provide the necessary information.

In addition, for both types of services we need the possibility to verify the produced evidence records. This can be done offline by a standalone application as well as by an online web service. Primarily in the web service scenario it is supposable that the rather complex process of verification is offered by the service as an additional functionality instead of burden the client with it.

The two kinds of services require a similar interaction protocol. The protocol must allow access to the evidence records and it must provide the possibility to initiate the process of securing the documents, either in notifying the service about new documents in a document repository or in sending the hash values of the documents.

At Fraunhofer SIT, we developed the software package ArchiSoft[1], which implements the above-mentioned concepts for the timely renewal of electronic signatures. In periodic intervals the validity of currently used algorithms is checked out. If an algorithm is threatened, the renewal process will automatically be launched. It offers an interface to integrate ArchiSoft in an arbitrary document management system. Furthermore ArchiSoft can act as a web service including the possibility for online verification of evidence records. Therefore, with ArchiSoft both kinds of long-term archive services can be realized.

Currently, neither proprietary protocols of document management systems nor standards for document management like the Open Document Management API (ODMA)[2] or the Open Archival Information System (OAIS) [14] support the handling with evidence records. The above mentioned LTAP aims at the interaction with long-term archive services but it is still in an early stage. For this reason, we specified the Trusted Archive Service Protocol (TASP), an archiving protocol that meets the requirements resulting from the two kinds of services. Besides operations for managing archived documents, the protocol supports access to and verification of evidence records as well as the initialization of the long-term security process for selected documents.

2.3. Security Suitability of Algorithms

Archiving services caring for the automatic signature renewal must be able to check the current security suitability of used signature algorithms. In Germany the Federal Network Agency (in German: Bundesnetzagentur) is the legally authorized institution for the valuation and annual publishing of the security suitability of signature algorithms. Since the publication is available in text form, it is not interpretable by machine or program.

Therefore a machine readable data structure has been defined holding the security suitability of algorithms and their parameters [3, 4, 17]. This structure is called a policy. It is already tried and field-tested both within the project ArchiSig and in the software ArchiSoft.

[1] http://www.sit.fraunhofer.de/archisoft

[2] http://www.infonuovo.com/odma/

A security suitability policy consists of policy name, different security suitabilities, validity period and the signature of the issuer. As seen in the following exemplary XML snip each security suitability is composed of the algorithm (algorithm identifiers, function, and parameters), validity start and end date.

```
<securitySuitability>
  <name>RSA</name>
  <algorithmType>asymmetric</algorithmType>
    <trustedAlgorithmOIDs>
      <OID>1.2.840.113549.1.1.5</OID>
      <OID>1.3.36.3.3.1.2</OID>
      <OID>1.2.840.113549.1.1.1</OID>
    </trustedAlgorithmOIDs>
  <trustedAlgorithmURIs>
    <URI>
      http://www.w3.org/2000/09/xmldsig#rsa-sha1
    </URI>
    <URI>
      http://www.w3.org/2001/04/xmldsig-more#rsa-ripemd160
    </URI>
  </trustedAlgorithmURIs>
  <function URI="http://www.archisig.de/id-RSAalgReq">
    <constraintValue>
      <RSAAlgConstraints modulusLen="768"/>
    </constraintValue>
  </function>
  <validityStart>1998-01-01</validityStart>
  <validityEnd>2000-12-31</validityEnd>
  <furtherInformation/>
</securitySuitability>
```

Up to now, the annually published security suitability policy has to be manually transformed from text document to this data structure by every user. This is an errorprone procedure. Additionally the administration and utilization of these policies expends effort.

For that reason a web service, which generates the machine readable policy and makes it available to the clients, could be an alternative. The created policy is electronically signed to protect it from manipulations and to enable authentication of the issuer.

Another web service may directly provide information about the security suitability of algorithms. Most important requests may be "Is an algorithm with explicit parameters currently suitable?", "Has an algorithm with explicit parameters been suitable at a specific time in the past?", "When is the validity of an algorithm with explicit parameters going to end?", "Until which date an algorithm with explicit parameters has being suitable?", "Which algorithms are suitable now?" and "Which algorithms will expire next?" The web service implements functions to perform such requests on the base of the policy and sends a response. In the process the web service has to care about the actuality of the security suitability policy. This alternative would relocate the preparation, updating and evaluation of the data structure from users to one trustful web service.

3. Legally Secure Transformation of Signed Documents

Solutions for the aging problem of electronic signatures are already in an advanced state, whereas the situation in the area of a secure transformation of signed documents is far from standard. First considerations concerning notary and certification services have been made in [18]. For the moment it's too early to speak of a state of the art technology in this field.

The German legislator has already passed respective laws that allow a certification in an electronic form. In particular these are the Administrative Procedures Law (in German: Verwaltungsverfahrensgesetz) and law concerning certifications by notaries (in German: Beurkundungsgesetz). Means combining organizational measures with technical solutions for legally secure transformations of signed documents are needed.

3.1. First Solution Concepts

The German Federal Ministry of Economics and Technology has funded the ongoing project TransiDoc[3] with the aim to develop concepts and solutions for the preservation of the probative value of transformed documents. TransiDoc has already delivered several results [2, 5], which are described in the following.

The primary concern of TransiDoc is that the probative value of the target document, i. e. the result of a transformation, is comparable to that of the source document. This is a prerequisite that the transformed document can be used in a later forensic argumentation instead of the source document. The presentation of the source must not be necessary.

The following ten principles defined by the project TransiDoc have to be met in order to achieve the abovementioned objectives of a secure transformation of signed documents. These requirements do not only address transformations between electronic documents (E→E), but also those involving paper documents as source or target (P→E and E→P):

1. Technical solutions are preferable; i. e. standardized technical proceedings should be used to mitigate human error and to increase (economic) efficiency.

2. Agreement of contents must be assured. The context of the respective use-case will determine the necessary degree of agreement resp. correlation between source and target contents. It may be required, for example, that errors and deficiencies of the source document, like invalid signatures, have to be recorded. In some transformation cases purposeful changes of the source contents are necessary, e. g. in order to make the contents anonymous.

3. Authentication of the authorship, i. e. the signers of the source documents must be authenticated. The authentication results including all used verification data, like certificates and revocation information, must be recorded. Handwritten signatures must be scanned in a careful way. If necessary, further information of the authorship has to be collected.

[3] http://www.transidoc.de

4. Data integrity of the source and target documents and of all protocols and verification results generated during the transformation has to be assured.

5. Attributability of the transformation, i. e. it must be verifiable who was responsible for the transformation. An electronic signature of an adequate security level will reach attributability of the transformation, if the result of the transformation is an electronic document. Otherwise a paper document will be hand signed, if required.

6. Authorization of the responsible persons, i. e. transformations should only be performed by authorized persons. It must be possible to verify the identity, role and authorization of the responsible persons later on. The respective use-case determines proper credentials, e. g. attribute certificates for an electronic signature. For a paper document the handwritten signature may be completed with an official seal. Institutions where transformations take place must develop own authorization concepts based on legal regulations.

7. Data protection and secrecy must be guaranteed, i. e. only authorized persons are allowed to access data used or generated during the transformation process. The personal data collected in order to verify an electronic signature must be restricted to the absolutely necessary amount.

8. Long-term usability of the target document, i. e. the source document should be converted in a long-term preservation format, which is unambiguously interpretable and standardized. This is the precondition for both the usability of the document over a long period of time and protection of the probative value of its content. Besides the converted content the target document contains protocols and verification results. These data must be structured in a way that it can be presented in various levels from giving an overview to going into the smallest detail. The presentation of the target document shall be adaptable to different purposes and users who may be specialists in this field or persons with less previous knowledge.

9. Traceability of the transformation, i. e. the complete workflow of a transformation, including information about the used systems and acting persons must be recorded to enable forensic evaluation. Evaluation must be possible independently of the transformation system.

10. Reliability of the transforming system, i. e. only certified and / or well audited systems should be used during the transformation process.

Considering the principles presented above it is obvious that it is not enough to just convert the content of a document to fulfill these principles. Even in the case of simple format conversions, additional process steps, like assay steps, and organizational measures are required to ensure security. To identify these requirements, a procedural analysis of a generic transformation process is helpful. In the following, a secure document transformation is presented as a sequence of high-level transformation phases. This phase model is generic, i. e. it is independent of the kind of transformation (E→E, P→E, or E→P) and it can be applied to a wide range of applications. Use-cases determine which phases are

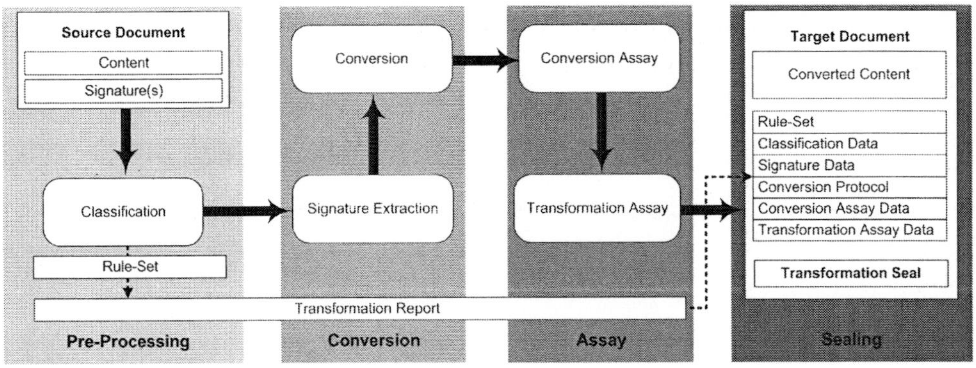

Figure 2. Procedural view on secure document transformations.

necessary, which one may be subsumed under or combined with another, or which may be parallelized if they become logically independent.

Figure 2 shows a secure transformation process starting from the source document consisting of its content and one or more signatures and resulting in the target document.

The intended purpose of a transformation influences especially the initial transformation phase called *classification*. The purpose may be to create a replacement, where the target shall be usable in the same way as the source document, such that, ideally, the source can be dispensed with. Another purpose may be to get a partial copy of the source usually containing only parts of the source content and accordingly limiting the possible usages of the target. In the classification phase the source document is inspected and those properties are determined which shall be recovered in the target document in order to fulfill the purpose of the transformation. That means, the classification does not only determine the relevant properties of the source but also those of the target and the complete transformation process, which have to be satisfied to achieve the desired reliability. From this classification follow the rules that govern the transformation process. The *rule-set* depends on every special application case, comprising organizational provisions, technical measures, attributions of responsibility, rules for signature verification and generation, etc. In general, the rule-set consists of a combination of machine-processable instructions, with normative prescriptions understandable only by humans. The *transformation report* is a data container which carries the information compiled during the transformation, such as protocols and assay results performed in later phases.

The real conversion is divided in two phases: *signature extraction* and *conversion*. During signature extraction, all relevant information of the source's signatures are gathered and added to the transformation report. The rule-set determines whether electronic signatures must be verified or the signers of the handwritten ones must be authenticated. The rule-set prescribes verification policies and decides the signature data to be carried to the report (e. g. time stamps, attributes etc.), and the result of the validation is also added to the report. In the conversion phase the proper conversion of source to target contents takes place according to the rules of the rule-set. Apart from the target contents, a conversion protocol and an error log are filed in the report. In the next phase the *conversion assay* follows, i. e. the correctness of the conversion is verified, e. g. by inspecting logs in the report and / or explicit comparisons of source and target contents. The final phase *transformation assay* serves two purposes; at first a consistency check on all preceding phases takes place, after that the result of the transformation with all data generated during transformation is secured. For this, a *transformation seal* is attached to the transformed document and signed by the transforming entity. The result of a transformation needs to be secure even if the source is not available for later comparison. This is the main reason why relevant data produced in the transformation process must be persistently and trustworthy bound to the target to enable a thorough forensic inspection. This possibility to assess the quality of a transformation is an important building block for the probative value of the target. It is embodied in three subordinate goals, which describe the essential purpose of the transformation seal:

1. Securing the integrity of the transformed document and other recorded data.

2. Attestation of the correctness of the transformation according to the specified rule-set.

3. Attribution of the transformation to the transforming entity and non-repudiation of that fact.

The result of a secure document transformation is the target document consisting of the converted content, the rule-set, the transformation report and the transformation seal. The project TransiDoc has developed data structures in XML notation for all parts of the target document. The data structures are as far as possible generic, i. e. application independent. They are flexible and extendable to support diverse certification services, architectures and different security levels. They enable the later verification of the transformation and provide means to prove the authorization of attesting authority.

The data structure of the rule-set is designed like a workflow definition, which determines how the transformation process shall be performed. The workflow definition consists of a sequence of activities. Each activity is either a manual or an automatic activity. A manual activity has to be executed by a human, whereas an automated activity shall be performed by a software or hardware component. The correspondent part of the workflow definition is the transformation report which provides what has actually been done during the transformation. For example the rule-set may require that a signature creation device conforming to the German Signature Law shall be used for creation of the transformation seal. The task of the transformation report is to provide the information about the actually used device, inclusive name, version and manufacturer. In detail, the transformation report consists of data processing protocols, documentation of the operations over the data, used software and hardware components and intermediate results, like signature verification results. It provides both traceability of data processing and authentication of the signer(s) of the source document. The transformation seal consists of an annotation and a signature. The annotation contains e. g. names of signers of the source document, including the authentication results, credentials of the certification service and its operators, time of certification and the attestation, e. g. of the correctness of transformation. Signatures contain the signature itself and e. g. certificates and attribute certificates. Hence the transformation seal provides the meaning and the trustworthiness of the assertion and data integrity and authenticity of the target document.

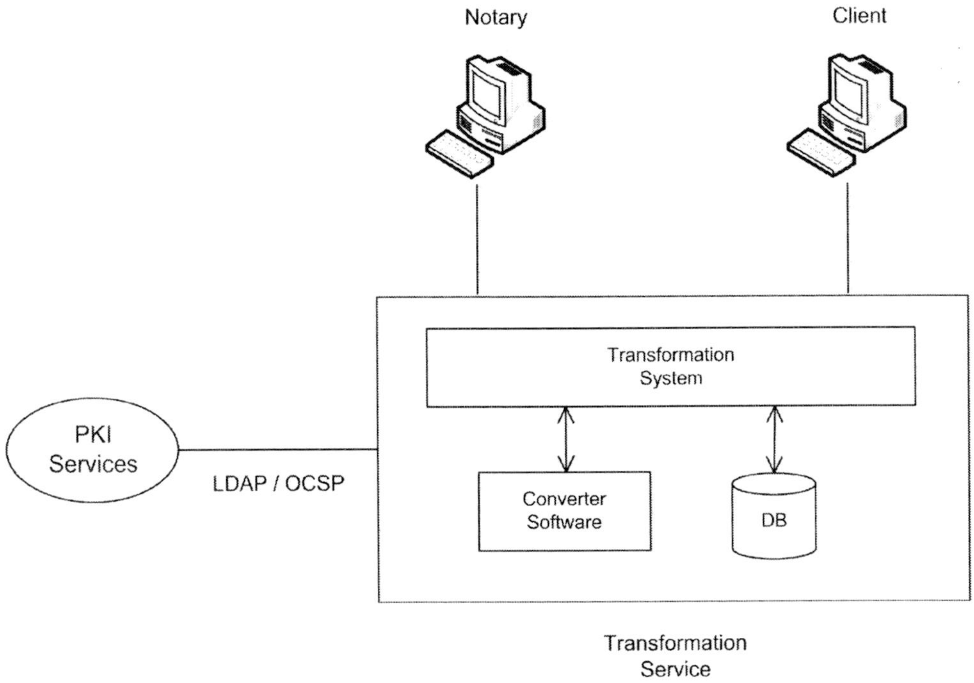

Figure 3. Transformation Service.

3.2. Transformation Services

Based on the introduced concepts and data structures, we propose a service for legally secure transformation of signed documents (see Figure 3). Examples are certification and notary services acting as a trusted third party who may attest that two documents contain the same information as well as the validity of the contained electronic signatures. Users can submit signed documents in order to let them transformed by the service into another data format without loss of their probative value. Note that different from paper-based documents, attestations of literal copies of digital documents only make little sense. In many cases the transformation process cannot be performed completely automatically. Only individuals can in general determine that the contents of two documents are equivalent. Another reason is that at least according to German law an attestation must be signed by means of a qualified signature, i. e. the signature of the resulting target document must be authorized by an individual. That means the service must allow the possibility of interventions by individuals. As a consequence, the service must support asynchronous communication since the user does not get the target document as an immediate response. After receiving a signed document from a user, the service performs the transformation according to the above mentioned transformation phases. After that, the service notifies the notary (e. g. via e-mail) so that he can perform the final transformation phase by verifying the correctness of the transformation and signing the target document. When the target document is completed, the user will be notified by the service so that he can fetch the target document.

Besides services for transforming documents, we need the possibility to verify target documents. This can be realized as a stand-alone application as well as an online service.

Verification of a target document means verifying the transformation seal and presenting the components of the target document (converted content, result of signature extraction etc.) in a comprehensible and retraceable manner.

To use this service, we need a rather simple protocol which allows the client to order a transformation, to query the status of the transformation, and to fetch the result of the transformation.

Both the transformation service as well as the verification service are currently developed by Fraunhofer SIT within the project TransiDoc.

4. Conclusions

We presented concepts and services for both main topics regarding long-term security: aging of signatures and changing of data formats. For both topics we have data structures which represent the performed processes and we have protocols for the interaction with the respective service. In principle the protocols could be combined to one protocol which covers all kinds of services introduced in this paper. But there is a difference in these concepts concerning the probative value. The probative value which can be obtained when using the concepts for the renewal of signatures is of another quality than the probative value achieved when transforming a signed document according to the introduced concepts.

The renewal of signatures by building hash trees and time stamping leads to a probative value that is based on cryptography. By means of the evidence record, the whole renewal process can be verified at any time. But the precondition of the security of this concept is that the signatures have always been renewed before the respective algorithms became insecure.

In contrast the probative value achieved when performing a legally secure transformation is primarily based on trust in the attestor or the service who transformed the document. Since in general, the source document is no longer available after the transformation, we cannot compare the original with the transformed document, and thus we cannot verify if the transformation is done correctly. We can only comprehend which steps were accomplished during the transformation process and – most important – we can verify whether the transformation seal was created by an authorized individual. For this reason, meeting the principles for legally secure transformations is essential to guarantee a correct transformation process. Furthermore, organizational instructions, e. g. concerning the security of the working environment, the choice of adequate technical security devices, or the individual who is allowed to perform the transformation, are of greater importance as with the process of signature renewal.

With this conclusion and the results presented in the sections before, we can now describe a service which cares for the timely renewal of signatures as well as for the transformation of documents. Because in general, document formats will change over the years, the service can automatically convert documents into subsequent data formats. It is noticeable that after a transformation, the existing evidence record cannot longer be verified since the hash value of the transformed document does not match to the one contained in the evidence record. Therefore, before transforming a document we must verify the evidence record together with the source document. The verification result must be recorded in the target document which is created when transforming the document. After that, we can delete the source document as well as the old evidence record. The target document will now be filed as a new data object and it will be time stamped and so, a new evidence record is generated. When we verify the new evidence record at a certain point of time in the future, we must also verify the target document, which means that we present the verification result concerning the old evidence record. As pointed out before, at this point of time, we can only trust in the correct transformation. If sometime in the future another transformation will be necessary, we will not only have to preserve the verification result of the last evidence record but also the verification results of all previous evidence records.

In order to automatically choose an appropriate data format, we can imagine a catalogue of data formats and their suitability for transformations. For example, some data formats should be avoided because of active contents they are not transformation-secure. Therefore requirements for data formats like long-term stability or platform independence must be established.

5. Summary

As we have shown, mature concepts for both the renewal of electronic signatures and the secure transformation of signed documents exist. Based on these concepts, we introduced different kinds of services which may help organizations to long-term and law-compliantly preserve electronic documents. In particular by the definition of web services SMEs and individuals gain the possibility to easily secure their documents over long periods of time without to be forced to deal with complex processes. This can contribute to encourage the application of electronic signatures.

Nevertheless, there are still some open issues. A prerequisite for the establishment of the evaluation of cryptographic algorithms and for their usage by national agencies (e. g. the Federal Network Agency) is that the proposed data structures will be standardized. Furthermore, it would be desirable to push on the standardization efforts for secure transformations.

References

[1] Brandner, R.; Pordesch, U.; Gondrom, T.: *Evidence Record Syntax (ERS), RFC* **4998**, August 2007.

[2] Fischer-Dieskau, S.; Kunz, T.; Schmidt, A. U.; Viebeg, U.: *Grundkonzepte rechtssicherer Transformationen signierter Dokumente*, Sicherheit 2005, GI-Edition, *Lecture Notes in Informatics* (LNI), Vol. P-62, pp. 401-412, 2005.

[3] Kunz, T.; Okunick, S.; Pordesch, U.: *Data Structure for Security Suitabilities of Cryptographic Algorithms (DSS),* Internet-Draft <draft-ietf-ltans-dssc-00.txt>, June 2007.

[4] Roßnagel, A.; Schmücker, P.: *Beweiskräftige elektronische Archivierung – Bieten elektronische Signaturen Rechtssicherheit?*, Economica Verlagsgruppe Hüthig Jehle Rehm GmbH Heidelberg, 2006.

[5] Schmidt, A. U.; Loebl, Z.: Legal Security for Transformations of Signed Documents: Fundamental Concepts, EuroPKI 2005, *Lecture Notes in Computer Science,* Vol. 3545, pp. 255-270, Springer-Verlag, 2005.

[6] Wallace, C.; Pordesch, U.; Brandner, R.: *Long-Term Archive Service Requirements*, RFC 4810, March 2007.

[7] Brandner, R.; Pordesch, U.: *Long-term conservation of provability of electronically signed documents*, ISSE Paris, October 2002.

[8] Merkle, R. C.: Protocols for Public Key Cryptosystems, *IEEE Symposium on Security and Privacy,* Oakland, CA, U.S.A., 1980.

[9] Jerman Blazic, A.; Sylvester P.; Wallace, C.: *Long-term Archive Protocol (LTAP)*, Internet-Draft <draft-ietf-ltans-ltap-05.txt>, July 2007.

[10] Directive 1999/93/EC of the European Parliament and of the Council December 1999 on a Community framework for electronic signatures, *Official Journal of the European Communities*, 19 January 2000, L13/12.

[11] Fischlin, M.; Pordesch, U.: *Nichtabstreitbarkeit trotz Verschlüsselung*, Datenschutz und Datensicherheit (DuD), Vol. 3/2004, pp. 163-168, Vieweg Verlag, 2004.

[12] Gudgin, M.; Hadley, M.; Mendelsohn, N.; Moreau, J.-J.; Nielsen, H. F.: *SOAP Version 1.2 Part 1: Messaging Framework*, W3C Recommendation, 24 June 2003.

[13] Chinnici, R.; Moreau, J.-J.; Ryman, A.; Weerawarana, S.: *Web Services Description Language (WSDL) Version 2.0 Part 1: Core Language*, W3C Candidate Recommendation, 27 March 2006.

[14] Consultative Committee for Space Data Systems: *Space data and information transfer systems – Reference Model for an Open Archival Information System (OAIS)*, Blue Book, Issue 1, CCSD, S2002, ISO 14721.

[15] European Telecommunications Standards Insitute: *ETSI TS 101 733 v1.5.1. – Electronic signature formats*, 2003.

[16] ISIS-MTT Specification, Preliminary Specification: *Long-term conservation of electronic signatures*, final draft, June 2004.

[17] Frye, F.; Pordesch, U.: *Berücksichtigung der Sicherheitseignung von Algorithmen qualifizierter Signaturen*, Datenschutz und Datensicherheit (DuD), Vol. 2/2003, pp. 73-78, Vieweg Verlag, 2003.

[18] Schmidt, A. U.; Gondrom, T.; Masinter, L.: *Requirements for Data Validation and Certification Services*, Internet Draft <draft-ietf-ltans-notareqs-03.txt>, December 2005.

In: Long-Term and Dynamical Aspects of Information... ISBN 1-60021-912-8
Editors: Andreas U. Schmidt et al, pp. 141-148 © 2007 Nova Science Publishers, Inc.

Chapter 11

PROVIDING EVIDENCE WITH TRANSFORMED SIGNED DOCUMENTS

Daniel Wilke*
University of Kassel
Project Group Constitution Compatible Technology Development

Abstract

The transformation of a document from one into another technical format is a central issue in electronic legal and commercial relations. Electronic documents are produced in many different software formats. For the sake of a practicable handling of these documents transformations are often necessary. Given the rapid development of the software market, such a change of the data format is essential because the readability of the documents could otherwise not be guaranteed. The need for transformation gives rise to new challenges both to technology design and the development of the law.

1. Introduction

Transformation can be effected in three different manners. It can be necessary to convert an electronic document from one software format (e.g. Word) into another (e.g. PDF). Furthermore, it can be necessary to transform a paper document into an electronic document by using a scanner. Equally, situations may arise where electronic documents need to be printed out and thus be transformed into a paper document.

In the life-cycle of a document that is to be archived for a long period, multiple transformations are almost inevitable. First, the original document often needs to be migrated into a suitable archiving format. Hence, where a paper document is to be introduced into an electronic archiving system, it has to be scanned and thus to be transformed into an electronic document. A similar procedure is needed if the original has been generated in electronic form from the start. For electronic documents, it is necessary to convert them into one of the software formats that have been chosen for the archiving system. Second, given the

*E-mail address: d.wilke@uni-kassel.de

rapid development of the software market, it will be essential to transform the electronically stored documents into new data formats before the old ones become obsolete. Otherwise it would be impossible to guarantee the permanent readability of the archived materials. Finally, it is by no means sure that in the near future everyone who wishes to use a document in legal relations will have adopted an entirely electronic workflow. Therefore, it will still be necessary to print electronic documents and transform them into paper documents in order to make them available to the reader.

Despite its obvious advantages the transformation of signed documents always involves the risk of falsification. Moreover, during the process of transformation the technical security devices lose their functional capability. Thus, the handwritten signature in a paper document loses its probative value when it is scanned. Similarly, once an electronic signature has been transformed into another software format the electronic signature cannot be verified in the target document which is generated in the transformation process, but exclusively in the source document. Hence, there is a need for suitable technical transformation systems which allow the preservation of a probative value in the target document that is comparable with the probative value of the source document. In the interest of cost-effective document management procedures it is desirable to create an automated transformation process and to design it in a way which makes the preservation of the source document unnecessary.

2. Technical and Organisational Measures for Legally Secure Transformations

Legally secure transformations of signed documents require technical and organisational measures that allow for a migration of the data and metadata of the documents concerned without putting at risk their probative value.

2.1. Guidelines for Proper Transformation Systems

If the law is to attribute a high probative value to transformed documents, legal practitioners need clear criteria to decide when a transformation is sufficiently secure to preserve the probative force of the source document in the target document. For this reason it is necessary to define a set of clear technical and organisational guidelines for proper transformations[1]. These should consider the following security measures.

The probative value of transformed electronic documents can only be augmented if the transformation process is carried out in a secure way which does not allow for any manipulation or loss of data. Hence, it is necessary to use reliable scanners, converters and printers in order to assure the correspondence of the source and target document. Furthermore the person who is responsible for the transformation process should attach a qualified electronic signature to the target document in order to guarantee the integrity of the target document. Moreover, the signature will allow the attribution of the transformation to the responsible person.

[1] For a draft see www.transidoc.de.

Another important element of legally secure transformation is to make sure that the transformation system is exclusively operated by authorised personnel. Hence, the institution carrying out the transformation should establish an infrastructure that denies access to the transformation system to unauthorised persons. Apart from such security measures, the user of a transformed target document should be enabled to control the transformer's authorisation. This could be implemented by means of attribute certificates which display the role (e.g. notary public, civil servant) of the person who has initiated the transformation process. Finally, to guarantee the long-term readability of the target document, it has to be generated in a stable standard data format.

2.2. Automated Transformation Processes

The creation of automated transformation processes is desirable for several reasons. As a matter of fact a lot of mistakes might be caused by human intervention. This risk could be avoided by establishing a fully automated process. Where such a system exists, the person who is responsible for the transformation could carry out subsequent random tests by comparing several source and target documents and thereby controlling their correspondence. Furthermore, automated transformation is a vital precondition for cost-effective document management. Many business companies, hospitals and administrative authorities preserve thousands of documents in their paper-based or digital archives. In these cases it would be very time-consuming and expensive if the correct transformation of every single document were to be controlled by a human being. Hence, a fully automated transformation process is needed.

As aforementioned, an electronic target document which is generated in the transformation process should be provided with the qualified electronic signature of the person responsible for the transformation. In a fully automated system, this signature would have to be created automatically as well. As long as the signature-creation device remains under the exclusive control of the holder of the private cryptographic key, this would not impede the validity of the signatures [1].

The system would have to verify automatically the unambiguousness and the signature of the source document as well as the correspondence of the source and the target document. In the first two cases the verification is a solely technical procedure so that it can be conducted automatically without difficulties [2]. The more critical aspect however, is whether the requirement of correspondence of the source and target document in form and content is reconcilable with automated transformation. Establishing the correspondence of two documents means controlling the result of a technical process. The system would have to literally compare every single character of the documents. This would not be feasible in an automated procedure. The control of the correspondence might seem expendable though, where the same result could be achieved by controlling whether the automated transformation has proceeded properly. The person responsible for the transformation would consequently be required to check the operability of the system at regular intervals and to arrange random tests concerning the correspondence.

2.3. Certification Infrastructure

Where a transformed document is used in legal proceedings, it would be rather cumbersome to assess the quality of the transformation process by retracing whether the guidelines for proper transformations have actually been observed. Therefore, it seems desirable to establish a certification infrastructure with a trustworthy authority auditing and certifying the compliance of the employed transformation system with the guidelines. The certification authority could either be a public institution like the German Federal Network Agency or a private service provider like the German TÜV Informationstechnik[2] (safety standards authority for information technology) endowed with sovereign powers to grant a certificate. Further information on the sovereign powers of safety standard authorities can be found in [3].

3. Certified Copies in the Digital World

In the paper environment a high probative value of paper copies is made possible through certification by a public authority or a notary public. Since the enactment of the Third Act Modifying Administrative Procedures[3], and the Judiciary Communications Act[4]. German law also provides for the certification of electronic documents which have another software format than the electronic original as well as printouts of electronic documents. The certification of transformed documents by a notary public has been thoroughly examined in [4]. However, these different kinds of certifications also serve different purposes. Certifying a paper copy serves as a means to multiply the original document so that the bearer can present it to several other persons or institutions without having to relinquish his original. After copying it he will still preserve this source document in his archive.

By contrast, scanning a document or converting it into another format serve a completely different purpose. Once an archivist has scanned a paper document and saved the hereby generated electronic document on a data medium, the latter can be copied and sent away as often as necessary. Nevertheless the archiving body does not have to dispose of its own electronic copy. The same applies where an electronic document is transformed into another format. As in these cases an electronic copy of the target document always remains with the archiving body, it is usually in its interest to destroy the paper or electronic source document in order to save storage space. Hence, in the case of format changing copies the certification aims at making the preservation of the original completely expendable.

Section 33 of the German Administrative Procedure Act defines legal requirements for the certification of transformed documents. The provision takes into account all three kinds of transformation. Having verified the correspondence of the source and target document, the competent public servant generates an attestation clause which he signs either by hand or electronically, depending on the form of the target document. Where the object of a certification is an electronic copy of an electronic document in another format, the target document must bear a qualified electronic signature.

[2]For further information on TÜV Informationstechnik see http://www.tuvit.de.

[3]Federal Law Gazette (BGBl.) 2002 Part I, p. 3322 et seq.

[4]Federal Law Gazette (BGBl.) 2005 Part I, p. 837 et seq.

The target document can only be an equivalent replacement for the source document if it attests that the latter had certain legally relevant properties, and that the target document corresponds with it. Such legally relevant properties are, for instance, the form of the document as well as information about the author's handwritten or electronic signature. Therefore, the certifying institution should attach an attestation clause to the target document which attests that the source document has these properties, and that the two documents correspond with each other. For the sake of legal certainty, the law should define how the integrity and authenticity of the source document are verified by the certifying institution and which verification data the attestation clause must contain. Section 33 (5) of the Administrative Procedure Act defines necessary verification data which the certifying person has to record in the attestation clause. Most importantly, these data are the name of the signature holder and the time when the signature of the source document was generated according to the signature verification. Finally, the attestation clause must state "which certificates with which data form the basis of the signature".

However, as long as the exact scope of these data remains unclear, it is doubtful whether this provision contains sufficiently clear work instructions for the certifying person [2]. If the probative value of the source document is to be preserved in the target document, the attestation clause must display detailed information about the signature of the source document. Thus, all certificates including the root certificate must be verified and the result of this verification must be recorded. Furthermore, it is important to incorporate information on the validity of the user's certificate at the time when the original signature was generated. This could be implemented by recording the result of an OCSP-request [5]. Moreover, it must be obvious from the target document whether the source document bore a qualified time stamp, as otherwise the validity of the certificate at the time of the signature creation could not be reliably proved. Finally, the attestation clause should establish if the hash and public key algorithms that have been used to generate the original signature were still sufficiently secure at the time of the transformation.

As the aforementioned verification data are of vital importance to retrace the transformation in case of a legal dispute, such an attestation clause or transformation seal should also be attached to target documents that have not been transformed by a notary public or a public authority.

4. The Probative Value of Electronic Documents

Electronic documents are, independent of the existence of an electronic signature, admissible as evidence in court. German procedural law provides five different means of proof. Apart from the interrogation of witnesses, expert witnesses and the parties themselves, the demonstrator can present deeds or submit any object to judicial inspection. Although both paper and electronic documents are means to store information, the latter are not legally regarded as deeds because they are not in written form and only readable by means of technical devices like a computer [6]. Therefore, electronic documents are a mere object to judicial inspection. As a basic principle these are subject to free estimation of evidence[5].

[5]Section 286 (1) Code of Civil Procedure.

In this case the judge can assess the evidential value of the document freely without being bound by any procedural rules. Where the genuineness of an unsigned electronic document is questioned, the party invoking its originality is responsible to proving it. In practice, the genuineness is made evident by means of a technical expert witness. The judge is generally free to appreciate whether or not to consider the document as genuine.

4.1. Original Electronic Documents

The principle of free estimation of evidence only applies if there are no statutory rules of evidence that are binding for the judge[6]. Such rules exist for electronically signed documents that are based on a qualified certificate. Thus, private electronic documents are presumed genuine unless there are justified doubts that the signed declaration originates from the holder of the private cryptographic key[7]. Public documents emanating from a public authority or a notary are presumed genuine unless the contrary is proven. In such a case the holder of the private key would have to prove that he has signed another text or no text at all or that someone else has created his signature unauthorisedly. Proving these defences in legal proceedings should be rather difficult [7]. Consequently, a public electronic document bearing a qualified signature enjoys a higher probative value than its private counterpart.

By enacting this rule, the legislator has followed the traditional distinction between private and public deeds. The reason for this distinction is that the original paper version of a public deed usually remains in legal custody, whereas the citizen only receives a duplicate [8]. Thus, the correspondence of the document presented by a party with the original can be easily verified[8]. Where the genuineness of an electronic document – private or public – is legally presumed its evidential value corresponds to the probative force of a deed[9]. Hence, private documents provide full evidence for the declarations that are contained in them[10]. Public documents, in contrast, provide full evidence for facts that are certified therein as well as for administrative decisions which the documents comprehend[11]. These rules are binding for the judge.

4.2. Transformed Electronic Documents

Although in the modern document management infrastructures transformations seem to be inevitable, some legal systems have so far not yet taken into account the practical need for replacing original documents. Under German procedural law, for instance, a private deed in paper form can only deploy its full probative force, if the original is presented to the judge[12]. Where a party presents only a copy of the document, the judge can consider the evidence freely without being bound to regard the declarations set down in the document as really executed. Public documents, by contrast, can also be presented in the form of certified

[6]Section 286 (2) Code of Civil Procedure.
[7]Section 371a (1) Code of Civil Procedure.
[8]Federal High Court of Justice, NJW 1980, p. 1048; Zöller/Geimer, ZPO, 24th ed. 2004, § 435, 2.
[9]Section 371a Code of Civil Procedure.
[10]Section 416 Code of Civil Procedure.
[11]Sections 415, 417, 418 Code of Civil Procedure.
[12]Section 420 Code of Civil Procedure.

copies[13]. However, even if the transformation system provides for maximum security with a human being controlling the correspondence of the source and target document, the latter can only be regarded as a certified copy of the original. This applies regardless of the kind of transformation concerned. Hence, the law in force impedes the use of transformed documents in judicial proceedings. This applies all the more where the transformation is carried out in an automated process without control of the correspondence of the involved documents.

Here again, the law of evidence privileges public document in comparison to private ones. This situation gives rise to the question of whether the different legal treatment of private and public documents is still adequate in the course of the changeover to electronic document management systems. Given the high level of security which is provided by electronic signatures, the integrity and authenticity of an electronic source document can be established with far greater reliability than in the case of a paper document bearing a handwritten signature. Where, moreover, a trustworthy institution like an administrative authority, a notary or a certified private organisation verifies these points as well as confirming the correspondence of the source and the target document, it seems appropriate to attribute a probative value to the target document which is comparable to that of the source document. If these requirements are met, it should not matter whether the document in question originates from a private person or a public authority.

5. Possible Modifications to the Law of Evidence

The participants in legal and commercial relations may hesitate to use transformed documents and to change over from paper based archives to electronic document management systems, as long as the evidential value of the transformed documents remains unclear. Under the law currently in force, even where guidelines on legally secure transformations are observed, the electronic target documents will be subject to the judge's free estimation of evidence, unless there are clear and binding rules defining their probative force.

Hence, the legislation could be adapted to the technological situation by creating a new legal provision which takes account of the security provided by modern transformation techniques. Although a target document that has been transformed in an automated process can hardly have the same evidential value as copy certified by a notary public or a public authority, it seems appropriate to accord it a higher degree of legal recognition. If a document has been transformed in the way described above and if it bears an attestation clause with a qualified signature there should be a rebuttable legal presumption concerning the accuracy of the statements in the attestation clause. Thus, the judge would have to presume that the transformation has been performed correctly, that the target document in hand corresponds with the source document and that the target document has not lost its integrity since the transformation. The presumption should not apply exclusively to documents that are certified by a notary public or a public authority, but to all documents, that have been properly transformed. The other party should remain free to prove the contrary [2].

[13]Section 435 Code of Civil Procedure.

6. Conclusion

Although under the existing technical conditions it is already possible to transform a document in a way which allows using the target document in legal relations, the concrete implementation of legally secure transformation concepts remains a demanding task. Apart from creating secure technical transformation systems it is also necessary to adapt the existing rules of evidence to the technical situation in order to preserve a high probative value in the newly generated document. If these measures are taken, the aim of legally secure automated transformation proceedings for large quantities of documents is reachable.

References

[1] Roßnagel, A.; Fischer-Dieskau, S.: *Automatisiert erzeugte elektronische Signaturen* MMR (MultiMedia und Recht), 2004, pp 133-139.

[2] Roßnagel, A; Fischer-Dieskau, S.; Wilke, D.: *Transformation von Dokumenten - Zur Notwendigkeit einer Technik- und Rechtsgestaltung* CR (Computer und Recht), 2005, pp 903-908.

[3] Maurer, H., *Allgemeines Verwaltungsrecht*; 14th ed., Verlag C.H.Beck, Munich, 2002.

[4] Malzer, H. M.: *Elektronische Beglaubigung und Medientransfer durch den Notar nach dem Justizkommunikationsgesetz* DNotZ (Deutsche Notar-Zeitschrift) 2006, pp 9-32.

[5] Fischer-Dieskau, S.: *Das elektronisch signierte Dokument als Mittel zur Beweissicherung,* Nomos Verlagsgesellschaft, Baden-Baden, 2006.

[6] Huber, M.: *§§ 371-494a ZPO.* In Kommentar zur Zivilprozessordnung; Musielak, H.-J.; 4th ed.; Verlag Franz Vahlen, Munich 2005; § 415, 5.

[7] Roßnagel, A.; Fischer-Dieskau, S.: *Elektronische Dokumente als Beweismittel - Neufassung der Beweisregelungen durch das Justizkommunikationsgesetz* NJW (Neue Juristische Wochenschrift) 2006, pp 806-808.

[8] Geimer, R.; *§ 435 ZPO.* In Zivilprozessordnung; Zöller, R.; 25th ed.; Verlag Otto Schmidt, Cologne 2005; § 435, 2.

In: Long-Term and Dynamical Aspects of Information... ISBN 1-60021-912-8
Editors: Andreas U. Schmidt et al, pp. 149-159 © 2007 Nova Science Publishers, Inc.

Chapter 12

DESIGN STRATEGIES FOR SECURE EMBEDDED NETWORKING

Axel Sikora*
University of Cooperative Education (UCE) Loerrach
D79539Loerrach, Germany
***Nathan Braun*† *and Steffen Jaeckel*‡**
Steinbeis Transfer Center Embedded Design and Networking at UCE Loerrach
D79539 Loerrach, Germany

Abstract

Albeit security is a major challenge for embedded systems in industrial, building and consumer applications, to-day most of the networked embedded devices run at a terribly low security level. This is not mainly due to the restricted resources on embedded systems, but appears to be a question of perception and knowledge of embedded designers. For the integration of security measures into embedded systems, a variety of additional requirements must be taken into account. Those are presented and discussed in this contribution with a special focus on embedded (inter)-networking. Selected solutions of the authors are presented, i.e. for embedded SSL.

1. Introduction

The integration of Internet connectivity and web services into embedded systems brings many potential advantages with it [1]. It combines two enabling technologies, both with low cost and high efficiency. It is crucial to understand that all use cases can be solved with legacy technologies as well. However, the major driver to use embedded web services lies in the huge potential to save cost, as they allow the integration of applications and communication into one device, the use of well known TCP/IP (and HTTP) Internet protocols with the low-cost and seamless integration with Ethernet, the quick, easy and solid development

*E-mail address: sikora@ba-loerrach.de
†E-mail address: braun@stzedn.de
‡E-mail address: jaeckel@stzedn.de

of HTML-based graphical user interfaces that can be deployed locally or remotely, and the use of standard web browsers for remote monitoring, diagnose and control.

However, a number of aspects should be regarded when using embedded systems in networks:

- Security is a major issue: Many embedded web servers may give access not only to data but may allow the control of devices, machines, and factories [5]. This results in the challenge to provide the highest possible security level at lowest cost [4].
- The risk of being attacked comes with the mere connectivity to the Internet [6] as there is a large number of systematic port scanners and as the data is transported via the public Internet. Therefore, the risk only partially depends on the application itself.
- Scalability is another issue. On the one hand, Internet protocols allow an optimum degree of scalability, as computing performance may vary from an 8 bit microcontroller to a 64 bit high end enterprise server without changing communication protocols or access mechanisms. Although the communication protocols remain the same for all these levels, the approach for security may differ significantly.
- Cryptography of course is a major means to tackle the security issue. The implementation of cryptography brings some peculiarities with it [3].
- The evolution of security threats and new means of deployment and therefore to provide possibilities to update an embedded system.

This contribution discusses the requirements and the general conditions of embedded network system (cf. ch. 2), presents practical solutions (cf. ch. 3), and demonstrates, how embedded systems can enable secure virtual private networking (VPN, cf. ch. 4).

2. Requirements and Challenges

2.1. Overview

This part describes the most important requirements, which must be considered when designing embedded systems for secure networking.

2.2. Basic Challenges

Of course, embedded systems mostly come with severely restricted resources. This may affect processing power and memory configuration. As this may be still acceptable for using symmetric cryptographic algorithms, calculation of asymmetric keys may become an extremely time-consuming issue. Poor resources make embedded systems extremely prone to denial-of-service attacks.

If the embedded systems has to work under real time conditions the popular approach of pure software solutions for cryptographic operations is not suitable.

2.3. Administration-Related Challenges

Many embedded systems come without any local user interface. Some even don't provide any button, as they are completely molded. This leads to the requirement that the complete administration has to be done via the network. Whereas this may be securely possible for wired networking, a huge challenge exists in case of wireless connectivity.

In many cases, embedded systems are delivered to the customer as ready-made systems. Whereas PC-users may be used to take care for their security, this is not the case for embedded systems. Product-specific root-passwords or master-keys are very wide-spread.

2.4. Long-Term Challenges

Many embedded systems are operated for months, years or even decades.

- In many cases, no re-boot of the system is acceptable. This puts extremely high demands to the stability of the system. The most critical part is the memory management, because it must be assured that it works absolutely stable in all conditions and after all events. Although this is an issue of safety, it is extremely related to security due to the challenging computational requirements of cryptographic algorithms.
- Most systems don't allow updates or patches, as boot-strapping from over the network is not supported. This means that today's systems run without any additions or adaptations, which requires a very future-proof design flow.
- The same holds true for master-keys held in the device, which cannot be changed on the device in case of no user-interface.

2.5. Application Support

In many cases, where embedded systems support sensor applications, its functionality is on the client side, as they should upload the data to a data sink, give an alarm or other status messages. In order to support management for those applications, a server side is preferable, as in this case the external system can master the management. Although there may be backdoors to this approach, it is closest to legacy PC-based architectures and poses minimal security risks.

However, this requirement implies the provision of a security protocol for both client and server sides.

3. Solutions

3.1. Overview

As the requirements from ch. 2 seem to pose quite high challenges to the system design and the system designers, quite often security measures are simply omitted. Most of today's embedded web servers run at an extremely low security level, thus opening a back-door to the complete system.

In order to address this issue, a variety of solutions are developed from the team of the authors. The objective of these stepping stones is to pave the way to a broader usage of secure systems.

3.2. Resource Management

It is of major importance that resource management remains stable for long-periods. This prohibits dynamic resource management, but calls for static task and memory management. This approach is quite common for all small embedded systems, however, may put an additional challenge to the overall system design flow.

3.3. Firewalls

In order to reduce the risk of denial-of-service attacks, firewalls are of major importance. For embedded systems, they come in two flavours.

- Application firewalls are strictly limiting the access to the system to the envisaged application. All other requests shall be ignored. This is the case especially when using a stand alone TCP/IP stack. Then all services resulting in open TCP or UDP ports are installed by the application programmer explicitly. In contrast, when using a more sophisticated operating system with pre-configured services, then applications itself but also in services that are configured wrongly or not configured at all and run with default settings.

- Stateful firewalls observe the frequency of attempts to access the system. In case of exceeding a given threshold, adaptive reactions must be triggered. In many cases, simple calculation of mean access frequencies may bring significant advances in security, without compromising the system's latency.

In terms of long term stability the lists and rules for the firewalls must be possible to be updated during run time.

3.4. Responsibility

Many embedded systems are not only observing, but also controlling industrial equipment. This leads to the risk that compromising the embedded system may also compromise the complete machine or factory. It is important to consider this issue, as a consequence may not only be a loss of information, which is bad enough, but a loss of control.

3.5. Embedded VPN

As already mentioned, cryptographic algorithms are a cornerstone of long-term security. For the embedded TCP/IP-stack from stzedn [8], two extensions are available, which seamlessly integrates the embedded device into an overall secure virtual-private networking (VPN) infrastructure.

- IPsec [2] allows static connections between systems and sites, mostly relying on symmetric cryptographic algorithms.

- SSL allows more dynamic connections between clients and servers, including hybrid cryptographic algorithms.

In order to relief the basic microcontroller from computational intense software operations, hardware co-processors can be used. They may be hardwired, as in newer microcontroller products, but can also be implemented in programmable logic [7]. The latter is especially attractive for soft-core microprocessors, e.g. NIOS [9] or Microblaze [10].

4. VPN on an Embedded System Using SSL

4.1. Strengths and Weaknesses of SSL for Embedded Systems

SSL-based VPNs promise many advantages to embedded systems, such as:

- The embedded system can be accessed with standard software such as a web browser.
- SSL allows client authentication without storing passwords on the embedded system. This is an essential benefit since embedded systems might operate in remote, inaccessible locations, where the exchange of secured data storage is not possible.
- SSL provides a variety of encryption standards that might be implemented mandatory. System designers find a balance between security risk, capabilities of the system and the strength of the encryption.
- SSL provides client as well as server authentication.

One major drawback of SSL proves to be the symmetric key generation that is computation intense and can not be easily and efficiently sourced out in hardware. On the other hand asymmetric encryption would increase the load during communication so that the drawback of the intense key generation must be taken into account.

The effort to provide client authentication must be considered carefully. Even though the authentication of clients on the server does not add a significant effort on the design, it requires the use of a PKI to distribute certificates to authorized clients. This effort has to be considered carefully when designing a system.

4.2. Design Considerations

To ensure long term security of an embedded system it must provide means of updating its firmware. Flash memories permit the update of certificates during run time.

In the same sense software might be updated as well. Ideally the application of the system allows a reboot so that a complete software update is possible, including also the adaptation of cryptographic algorithms.

In terms of long term security the use of configurable hardware such as Field Programmable Gate Array (FPGA) is the best choice. This makes a scalable hardware - software design possible. The use of encryption/decryption functions in hardware relieves the CPU significantly. Renowned manufacturers of FPGAs with optimized 32 bit CPU cores are Altera with its NIOS II platform [9] and Xilinx [10].

The drawback of FPGAs in comparison with standard microcontrollers is increased power consumption and higher prices per logic function (which might be alleviated through monolithic system integration), an increased complexity and the need for hardware synthesis tools.

As a result for a major number of embedded applications the choice lies still at standard microcontrollers. The following trends among standard microcontrollers can be observed:

- increase of performance: 32 bit microcontrollers are more and more used with internal CPU frequency in the range of 50 MHz and higher.
- increase of monolithic integration of communication peripherals such as Ethernet MAC and PHYs.
- increase of monolithic integration of hardware support for crypto algorithms such as AES, RSA, DES/3DES, or random number generation [11].
- increasing memory capacities on the chip.

The requirements to the software in regards of the still limited resources of embedded systems and of the vast variety of platforms can be seen in:

- efficient implementation in terms of execution speed and memory usage.
- portability can be achieved by a smart software interface design.

Portability with the definition of software interfaces unfortunately contradicts with efficient implementations. However, only portable (and thus reusable) software can be maintained at reasonable costs.

The implementation of the crypto wrapper is of high importance. It provides the means of generating target specific code. The network security layer must be able to access the same functions on different platforms. The implementation of the crypto library, however, may vary from one platform to another. If a platform provides hardware support then the crypto wrapper is configured to access those functions in order to offload the CPU. Only the functions that are not available in hardware are then run completely in software. Furthermore this enables the implementation of time critical functions in CPU optimized code such as assembly language.

4.3. Client and Server Side

The main differences between the Client- and Server-side can only be noticed while the handshake. The design aim of the SSL protocol was to enable a secure connection for Clients, to make it possible to verify a connection to one certain server, so the client makes its proposals of supported cipher suites and the server chooses that one that fits best to him. The specification of the SSL protocol defines a special way of permuting the given parameters which is done equally on both sides, server and client, with the same parameters, because the client has to know the servers key and the other way round. That will lead to the conclusion that key calculation is the same on both sides, simply the "other" part is used as

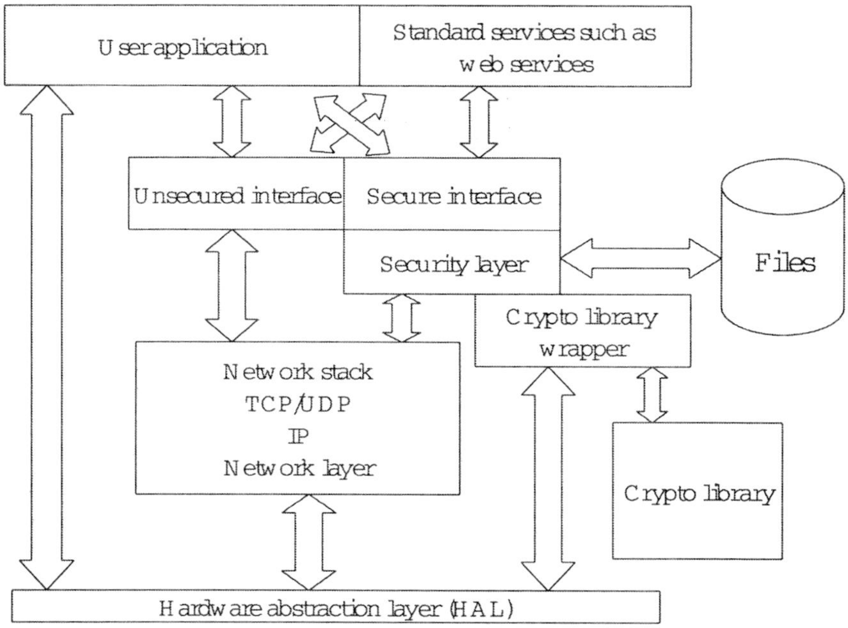

Figure 1. Software partitioning for a portable cryptography network stack.

symmetric ciphers' parameter for de/encryption, and there's no difference in performance when calculating the keys. When regarding performance issues there are two critical points. The first one always happens and appears in the ClientKeyExchange, the second one only appears when using mutual authentication and happens at the CertificateVerify message. In both cases RSA operations have to be done which costs a lot of effort and shouldn't be interrupted because mostly high-speed routines are implemented that work directly with registers and will crash on interrupt-return. The routines that are used to execute the RSA operations are oppositional in its jobs. When creating the ClientKeyExchange the client has to encrypt the message with the servers' public key and the server encrypts the message with its private key. In the CertificateVerify the client signs the message with its private key and the server can verify this signature with the clients' public key. After completion of the handshake both sides are treated equal, they have the same effort to handle the de/encryption operations and are for example always allowed to shut down a connection.

4.4. HTTPS on a Soft Core Processor

[8] has developped an SSL module as an extension to its emBetter TCP/IP stack for 8/16/32 bit microcontrollers. Its cryptographic extensions provide features such as:

- cipher suites AES128/256 and RC4 for symmetric encryption using RSA for asymmetric key exchange. ECC is under preparation
- client and server authentication
- fully non-blocking implementation to use with or without an operating system

- session resumption
- scalability in terms of memory usage

In the first run, the SSL module was implemented on a soft-core microcontroller:

- Altera Cyclone II FPGA
- 50 MHz NIOS II RISC processor with data and instruction cache
- 16Mbytes of program and data memory
- integrated 10/100 Mbit/s Ethernet MAC

The focus on this implementation was not only to achieve a high level of security but also a high data throughput. First measures delivered a decryption speed of some 40 kBytes/s for a TCP stream, without any optimization. Unencrypted data might be sent on the same system at around 4 Mbytes/s. This shows clearly the major effort of encryption technologies on embedded systems.

Further implementations on other low-end 32-bit microcontroller, e.g. ARM7/ARM9, are available.

4.5. Hardware Acceleration

In order to leverage the encryption and decryption speed, hardware acceleration is a major means. The following possibilities are at hand:

- use external dedicated coprocessors. This approach is always feasible, but mostly suffers from the time consuming copying of data to and from the coprocessor.
- use external chip-cards. Some systems come with a local chip-card based scheme for authentication. In these cases, the resources of the chip-cards may be used. Although the copying efforts remain, advantages can be realized for low data volumes and highly complex computation, as they are required for key generation via RSA.
- use internal hardwired crypto-coprocessors. More and more microcontrollers dedicated for networking applications are equipped with hardware accelerators, e.g Atmel AT91SAM9xxx
- use internal softcore crypto-coprocessors, as they are feasible in PLD-based systems. [14] shows some data from a realization for a 3DES demo system from the same authors.

4.6. Design Support

Economy of mechanism is one important design parameter to enable user acceptance. Therefore, it is of major importance that the integration and the configuration of the SSL functionality can be easily used. In order to support the application engineer in this task, an easy to use script called XCA [15] is used to handle the Public Key Infrastructure.

5. Long-Term Security of Embedded SSL

5.1. Memory Management

The memory management of the emBetter suite is completely kept static. This is an absolute requirement to all microprocessor systems without a memory management unit (MMU). As a consequence, memory fragmentation does not become an issue.

In addition, care must be taken concerning the dynamic execution of program code, i.e. stack behavior. Whereas this issue can be handled with a reasonable margin, the co-operation with the application software tends to be more error-prone. No single solution can be provided, except extensive testing (cf. ch. 5.5).

5.2. Mutual Authentication

In legacy computing systems, e.g. business IT systems, servers communicate with many clients. One-side authentication seems to be sufficient for the computing devices, as authentication can be ensured by user related credentials. In addition, as servers can be physically secured (cf. ch. 5.3), user-related information may be stored on them.

In embedded computing, machine-to-machine communication is much more relevant. Thus, devices itself must be authenticated. This should be done on a certificate level. Mutual certificate-based authentication is supported by emBetterSSL.

5.3. Physical Security

Legacy computing systems are separate machines, which can be centralized and secured in dedicated premises, whereas embedded systems are applied in their final system - and thus cannot be mechanically secured with major efforts to keep cost at a reasonable level. Major means to ensure the integrity of the systems are:

- use of microcontrollers with internal flash memory,
- use of internal flash memory, which can be secured against unauthorized read-out,
- use of additional physical measures to secure the microcontroller or an attached chip-card.

It should be clear that all of these measures can be tempered at the attacker's expense. emBetterSSL comes with different solutions to this issue. However, they strongly depend on the target system and application.

5.4. Update

Running machines for years and decades also implies the necessity of updates. They must be secured so that

- the update can be accessed only after a successful mutual authentication,
- the update is stable, before the older version is removed.

emBetterSSL may use the mutually authenticated SSL-secured communication channel via http file upload. Security against loss of data or interrupt of flashing process is achieved via flash mirroring, i.e. running double program memory.

5.5. Testing

Albeit theoretical analyses are the fundament for a reasonable long-term stability and security, testing remains a major practical means to find weaknesses and loopholes. emBetterSSL comes with an extended test-suite covering the functional use cases. In addition, further experience is to be collected with port-scanner like Nessus [16]. Due to non-standard implementation and no OS-related loopholes, the results are very relaxing. However, one should not forget about system specific weaknesses.

6. Summary

Long term security for embedded systems is closely related with the use of encryption technologies. The use of standard protocols vs. proprietary protocols promises the use of standard software on remote systems and thus cost reduction on the overall system.

On the other hand there is no standard hardware platform on an embedded system and the variety of implementations is big. In some cases the embedded system must provide a continuous uptime so that the update of software is only partially possible.

With embedded systems more and more interconnected implementing VPN's and fire walls must be a main focus in the design phase of an embedded system with Internet connectivity.

The use of SSL or TLS on embedded systems is of special interest since it does not require passwords stored within the embedded device.

Even though SSL requires significant resources embedded systems are capable to apply it due to increased processor performance, the break through of 32bit platforms and the implementation of hardware accelerators such as crypto co-processors.

References

[1] Sikora, A.; Bruegger, P.: Virtual Private Infrastructure - An Industry Consortium for Unified and Secure Web Control with Embedded Devices. In: *Proc. 9th IEEE International Conference on Emerging Technologies and Factory Automation (ETFA 2003)*, Lisbon, 2003.

[2] Riedmueller, S.; Brecht, U.; Sikora, A.: IPsec for Embedded Systems, in: H. Weghorn (Ed.), *Proceedings of the 2ndAnnual Meeting on Information Technology and Computer Science at the BA-University of Cooperative Education*, ITCS 2005, Stuttgart, 3. Mai 2005.

[3] Straub, T.; Sikora, A.: *Cryptography on Embedded Systems*, embedded world 2005 Conference, Nürnberg, S. 517-536.

[4] Sikora, A.: *Embedded Security for Ambient Intelligence*, VDE-Kongress 2004, Berlin, 2004.

[5] Byres, E.; Lowe, J.: *Real world cyber security risks for industrial control systems*, The Industrial Ethernet Book, September 2004, pp. 30-33.

[6] Eichin, M.: *Security Issues in Embedded Networking*, Embedded Systems Conference 1993, Apr. 1994.

[7] Prasanna, V. K.; Dandalis, A.: *FPGA-based Cryptography for Internet Security*, available at:
http://ceng.usc.edu/~prasanna/papers/dandalisOSEE00.pdf

[8] http://www.embetter.de

[9] http://www.altera.com

[10] http://www.xilinx.com

[11] http://www.atmel.com/dyn/resources/prod_documents/doc1593.pdf

[12] http://www.atmel.com

[13] *Wakan Crypto-Software-Toolkit Release 2.1 Technical Information*, Information at http://www.stzedn.de

[14] Braun, N.; Sikora, A.: *Design Strategies for Secure Embedded Networking*, Workshop *Long-term Security*, Emerging Trends in Information and Communication Security (ETRICS'06), 6.- 9. Juni 2006, Freiburg, Germany.

[15] http://sourceforge.net/projects/xca

[16] http://www.nessus.org/

INDEX

C

D

E

F

G

T

U

V

W

X

Y